The Readable Maths and Statistics Book

For
Andrew and Geoffrey

The Readable Maths and Statistics Book

BARRY EDWARDS
Senior Lecturer, Middlesex Polytechnic

London
GEORGE ALLEN & UNWIN
Boston   Sydney

First published in 1980

This book is copyright under the Berne Convention. All rights are reserved. Apart from any fair dealing for the purpose of private study, research, criticism or review, as permitted under the Copyright Act, 1956, no part of this publication may be reproduced, stored in a retrieval system, or transmitted, in any form or by any means, electronic, electrical, chemical, mechanical, optical, photocopying, recording or otherwise, without the prior permission of the copyright owner. Enquiries should be sent to the publishers at the undermentioned address:

GEORGE ALLEN & UNWIN LTD
40 Museum Street, London WC1A 1LU

© Barry Edwards, 1980

British Library Cataloguing in Publication Data

Edwards, Barry
   The readable maths and statistics book.
   1. Business mathematics
   I. Title
510'.2'4658   HF5691   80-40629

ISBN 0-04-31007-4

Typeset in 10 on 12 point Press Roman at the
Alden Press Oxford London and Northampton
Printed and bound in Great Britain by
William Clowes (Beccles) Limited,
Beccles and London

Contents

		page	
Preface			xi
1	**The Straight Line**		1
	Introduction		1
	Points and Lines		1
	Break-Even Analysis		5
2	**Linear Equations and Inequalities**		10
	Systems of Linear Equations		10
	Inequalities		12
3	**Linear Programming**		17
4	**Differential Calculus**		29
	Functions		29
	Differentiation		31
	Economics Applications		40
	Elasticity		43
	Differentiation of the Exponential Function e^x		45
5	**Partial Differential Calculus**		50
	Differentiation Revision		50
	Partial Differentiation		51
	Maximum and Minimum of Functions of Two Variables		61
	Use of Lagrange Multiplier		66
	Further Examples of Use of Lagrange Multiplier		69
6	**Integral Calculus**		75
	Integration		75
	Economics Applications		84
	Integration of the Exponential Function e^x		91
	Integration as the Area under a Curve and the Limit of a Sum		91
7	**Probability**		96
8	**Sets**		103
9	**Introduction to Statistics**		115
	Definitions		115
	The Frequency Distribution		116

	Graphical Presentation of Data	*page* 117
	Median and Mode	122
10	**Measures of Location and Dispersion**	124
	Summation Notation	124
	Measures of Location	125
	Measures of Dispersion	130
11	**Index Numbers**	137
12	**Correlation and Regression**	148
	Measuring the Degree of Association	148
	Regression Lines	152
	Coefficient of Determination	158
13	**Permutations and Combinations**	162
	Permutations	162
	Combinations	164
14	**The Binomial and Poisson Distributions**	167
	The Binomial Distribution	167
	The Poisson Distribution	174
15	**The Normal Distribution**	181
16	**The Probability Density and Cumulative Distribution Functions**	192
	Revision: Mean and Variance	192
	The Probability Density Function	194
	The Cumulative Distribution Function	196
17	**Samples and the Standard Error**	203
	Mean and Variance of Sample Means (Infinite Population)	203
	Mean of Sample Variances (Infinite Population)	205
	Variance of Sample Means and Standard Error	207
	Variance of Sample Means (Finite Population)	213
18	**Confidence and Further Distributions**	217
	Confidence Intervals and Levels	217
	Sum and Difference Distributions	220
	Proportions	224
19	**Hypothesis Testing**	229
	One-Tail and Two-Tail Tests	229
	Type I and Type II Errors	234
	Calculation of β (the Probability of a Type II Error) and Operating Characteristic Curves	235
	Test for Difference of Means	238
	Test for Difference of Proportions	240

20	**Small Samples and the t Distribution**	page 245
	The Student t Distribution	245
	Confidence Intervals	246
	Test for Means	248
	Test for Difference of Means	250
	Paired Samples	253
21	**The χ^2 Distribution and Its Uses**	258
	The χ^2 Distribution	258
	Goodness of Fit	262
	Contingency Tables	269
22	**Tests for Variance and the F Distribution**	275
	Tests for Variance	275
	The F Distribution	280
23	**Analysis of Variance**	286
	One-Way Analysis of Variance	286
	Two-Way Analysis of Variance	296
	Design of Experiments	300
	Conversion of ANOVA Formulae	301
24	**Sampling**	305
	Importance and Purpose of Sampling	305
	Types of Sampling	307
	The Effect of Stratifying a Sample on Standard Error	310

Appendix: Procedure Summary Sheets (Tests and Confidence Intervals)   315

Answers to Exercises   320

Index   326

Preface

The idea for this book came originally from the demand by some of my students for one volume that covered all the mathematics and statistics for their BA(Hons) Business Studies course at Middlesex Polytechnic. It had become increasingly clear to me over several years that the books that were available covered only specific sections in unnecessary depth. At the same time courses such as those for the Institute of Works Managers Certificate and Diploma would be catered for. When the book was finished, my colleagues teaching on social science courses advised me that their own students also needed this level of mathematics and statistics. Thus, the original concept widened.

Since almost every type of course relies so much on mathematics and statistics as a tool, it is hoped that everyone at universities, polytechnics and colleges, from accountants to engineers, will find much that is useful.

The statistical tables to which I refer throughout the book are to be found in J. Murdoch and J.A. Barnes, *Statistical Tables* for Science, Engineering, Management and Business Studies, 2nd edition (London: Macmillan, 1974). Another set of tables recommended as comprehensive and new is H.R. Neave, Elementary Statistics Tables (London: George Allen & Unwin, 1980).

My grateful thanks are due to Mrs Jean Hoy for her invaluable assistance with manuscript preparation. A considerable amount of inspiration and practical help came from Miss Tricia Doran, to whom I give my sincere thanks. I am also grateful to Lyn Kendal-Archer for her great help with the proofs.

BARRY EDWARDS
Stanford-le-Hope
September 1979

1 The Straight Line

INTRODUCTION

The subjects that seem to me to frighten students most are mathematics and statistics, frequently met in the guise of quantitative methods. This fear arises possibly because of the apparent multitude of impenetrable formulae, or perhaps because there is something chilling about a deceptive simplicity that often melts away to leave only riddles. Students cannot bring themselves to believe that the subject of mathematics, which is so fundamental to most other subjects, is really simple.

Consequently, I have tried to write in a particular readable style and to show, in presenting each new idea by means of a developed example, that there is an underlying unity of approach and method. Hopefully, this makes the concepts straightforward and the formulae redundant. For example, most of the statistical tests are not many separate objects but simply variations on one theme. What is necessary, but difficult to achieve after at least eighteen years of 'thinking complicated', is to work from first principles, to make no assumptions about any situation and to 'think simple'.

POINTS AND LINES

Our first task is to construct a system of names and conventions (called a *notation*) to describe the position of an object (which we may consider in the abstract as the location of a point). How do we describe to another person where something is or how to get there?

It is probably as well to realise immediately that locations, like almost everything else, are not absolute. Rather, they are relative to somewhere else, so where do we measure from? Where does the world start? Well, zero longitude for example when measuring positions round the globe is considered from Greenwich, but this is due to historic reasons. It is so easy to consider everything only from one's own personal or national viewpoint that I cannot stress too much how important it is to adopt a detached 'assuming nothing' approach to seek objectivity about a problem.

Let us consider then a system of location based on a two-dimensional grid with an arbitrary starting point or zero (Fig. 1.1) which you can think of as being like Fig. 1.2 except that the geographical direction does not have to be adhered to and it is convenient to label the two central grid lines or axes as shown.

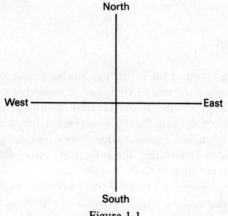
Figure 1.1

This is usually called the *Cartesian System*, after Rene Descartes the famous French Mathematician, and could if necessary be extended to three dimensions by using a box grid.

Where is A? Relative to O it is 1 unit east and two units north so we refer to it as the point (1,2). Notice we choose to always put x first and y second. The units are of course arbitrary; there is nothing magical about

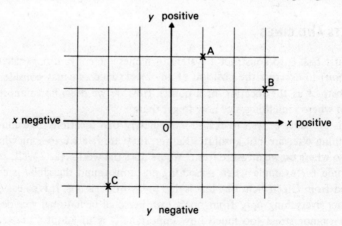
Figure 1.2

yards, metres or any other unit. Similarly B is at the point $(3, 1)$ and C is at $(-2, -2)$, both relative to O. If we were to choose C as our arbitrary 'world centre', and impose a similar system on the same layout as in Fig. 1.2, then the position of A in the new (X, Y) system would be $(3, 4)$. Where is B relative to A? I hope you can see it is $(2, -1)$.

Normally of course having chosen our arbitrary zero we keep to it.

We can now move on, using the notation we have just introduced, to consider the idea of a *linear relation*, something very important in the basic work you will do in several subjects. Suppose you have a brother three years younger than yourself with, for the sake of convenience, the same birthday. Obviously your age, which we can call y, and his age, which we can call x, are *variables* because time does not stand still and they vary. How are they related?

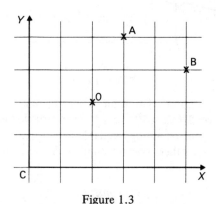

Figure 1.3

Whatever age x he is, your age y will be that and three years more, because you obviously age at the same rate. So, the linear relation between your ages is

$$y = x + 3$$

If your names were Andrew and Barry respectively, this could be written

$$a = b + 3$$

which would be absolutely equivalent.

Clearly, the relation holds for all possible ages (except before he was born), and we can consider particular cases in order to plot a picture of the relation on our Cartesian system. For example, when he is 1 year old, you are 4; that is, when $x = 1$, $y = 4$, which is really a point $(1, 4)$ such as we considered before. When he is 2 you are 5; when he is 3, you are 6; and so on to get Figure 1.4. We could plot hundreds more points, but they would all lie on, in fact create, this same straight line.

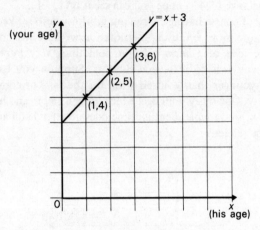

Figure 1.4.

The essence of a linear relationship is that there are no powers of x or y other than power 1 (e.g. $x^{-1/2}$ or y^3). One variable (y) is changing in a way that is dependent on another variable (x) so that y is equal to a constant number plus another constant number times x. The general way to write this is

$$y = mx + c$$

where m and c are the constants, y is called the *dependent variable* and x is called the *independent variable*.

Constant c (called the intercept) is the distance cut off by the line on the axis of y. In our example $c = 3$ (i.e. your age when your brother was born).

Constant m (called the *gradient*) is the slope of the line; that is, it measures the *rate of change* of y with respect to x, or how much y changes for each unit of change in x. Just like the gradient of a hill, it measures the ratio of the distance gone up vertically to the corresponding distance horizontally. In our example $m = 1$.

In other examples m and c can have any values. Consider the annual cost y of running a car, which will have a fixed part due to road tax and insurance, say £150, before you even drive it. It will also have a cost due to petrol and maintenance, which will, as a good approximation, vary directly according to your mileage x, say 8p per mile. So, we have (in pence)

$$y = 15{,}000 + 8x$$

or

$$A = 15{,}000 + 8M$$

or the same equation with any symbols that you want to use.

Consider the relation

$$8x + 4y - 3 = 0$$

Is this a straight line? Yes, because only powers of 1 are present, and we can easily rearrange it to our more convenient form as follows:

$$8x + 4y - 3 = 0$$
$$4y = -8x + 3$$
$$y = -2x + \tfrac{3}{4}$$

It is a straight line with gradient -2 and intercept $\tfrac{3}{4}$.

BREAK-EVEN ANALYSIS

This is a suitable point at which to introduce an application of linear relations to a topic met in accounting. Consider a situation in which a factory produces articles with both a fixed cost of production and a variable cost.

The fixed cost is due to costs in the initial setting up of the line, tool costs, market research prior to launching the product, and so on. Obviously, this is laid out initially with no guarantee of any sales at all but in the knowledge that it will not be added to and will effectively be shared out over the total production, giving an increasingly smaller share per article, the larger is the production run.

The variable cost varies directly (i.e. linearly) with the number produced and is due to actual material costs, labour costs, and so on. Clearly, ten times the number of articles use up ten times the amount of steel or whatever.

The basic idea is to discover how many articles must be produced so that the factory is generating enough total revenue TR (which also is likely to be linearly related to the number sold) to cover its total cost TC (i.e. to break even). At the break-even point

$$TC = TR$$

There are a number of important assumptions that we have to make in this model, apart from the assumed linearity of the cost and revenue. These are that:

(1) one product is produced;
(2) the production is all sold;
(3) costs are charged against revenue; and
(4) there is no change in prices or costs.

This is obviously a simple and limited model, but it will be illustrative, and you can obviously run on to more complicated models when you have learned to walk with this one.

Suppose that the total fixed cost of setting up a run of articles is £500 and the variable cost is 30p per article. They sell at 50p each. What production figure will put the firm on the threshold of profit? We construct the relations as before; hence,

$$TC = 500 + \tfrac{3}{10}x$$
$$TR = \tfrac{1}{2}x$$

where x is the number produced and the units are pounds. Break-even point in this case is where

$$TC = TR$$
$$500 + \tfrac{3}{10}x = \tfrac{1}{2}x$$
$$500 = \tfrac{1}{5}x$$
$$x = 2{,}500 \text{ articles}$$

We could show the same theory by plotting each line on a graph (see Figure 1.5). The point of intersection of the two lines is where they are both satisfied by the same values of x and y, which is merely a picture of the solution above. It shows clearly that, if x is less than 2,500, there is a loss, as TC is above TR; if x is above 2,500, there is a profit, as TR is above TC.

If production is 3,000 articles,

$$TR = £\tfrac{1}{2}\,(3{,}000) = £1{,}500$$
$$TC = £500 + £\tfrac{3}{10}\,(3{,}000) = £1{,}400$$

That is, a profit of £100, which is equal to the vertical gap between the lines at $x = 3{,}000$ on the graph, is obtained. If production is 2,000 articles,

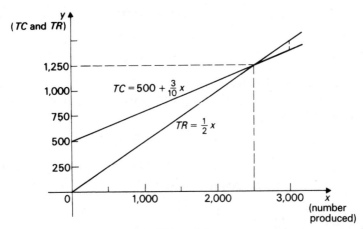

Figure 1.5

$$TR = £\tfrac{1}{2}(2{,}000) = £1{,}000$$
$$TC = £500 + £\tfrac{3}{10}(2{,}000) = £1{,}100$$

and a loss of £100 is made.

Let me now introduce one or two more names and confirm some ideas using the example (in pounds)

$$TC = 500 + \tfrac{3}{10}x$$

At a production level of 2,000 articles

$$\text{Total variable cost} = £\tfrac{3}{10}(2{,}000) = £600$$

since the variable cost per article is $£\tfrac{3}{10}$ or 30p. This last figure, which is really the gradient of the line, is also called the *marginal cost*, being the cost of producing one more article. *Average cost* per article takes into account fixed cost also, as well as the number produced:

$$\text{Average cost per article} = \frac{TC}{x}$$
$$= \frac{£1{,}100}{2{,}000} = £\tfrac{11}{20} \text{ or } 55p$$

at a production level of 2,000 articles.

Sometimes you may be given an example where selling price is not given but total sales value (i.e. revenue), total variable cost and total fixed cost are given instead. Say at some production level that

$$TR = £900$$
$$\text{Total fixed cost} = £100$$
$$\text{Total variable cost} = £400$$

It is then better to have the gradient expressed not as the variable cost per unit but instead as the variable cost per £1 of sales. Then, for this example we can write (in pounds)

$$\left. \begin{array}{l} TC = 100 + m\,900 \\ TC = 100 + 400 \end{array} \right\}$$
$$\therefore \quad 900\,m = 400$$
$$m = \tfrac{4}{9}$$

Substituting gives
$$TC = 100 + \tfrac{4}{9} TR$$

For break-even,
$$TC = TR$$
$$\therefore \quad TR = 100 + \tfrac{4}{9} TR$$
$$\tfrac{5}{9} TR = 100$$
$$TR = \tfrac{9}{5}(100)$$
$$= 180 \text{ (i.e. £180)}$$

for break-even. Hence, the profit or loss at a particular revenue level can be found.

EXERCISES

(1) Plot the following lines:

    (a) $y = 2 + 3x$ from $x = -3$ to $x = +3$.
    (b) $2y = 8 - 5x$ from $x = -1$ to $x = 5$.
    (c) $y = 2x$ from $x = -2$ to $x = +4$.
    (d) $-2x + 3y + 6 = 0$ from $x = 0$ to $x = 6$.

(2) A salesman's weekly earnings consist of a basic of £40 plus 20p in the pound commission. Construct the equation describing his earnings, and plot it on graph paper.
What will he earn on sales of £240?
How much must he sell to earn £100?

(3) A company has fixed costs of £9,000 and variable costs of £60 per item. Each item sells for £75. Find the number of items that must be produced to break even.

(4) The selling price of an article is £2. At a certain production level the total variable costs are £50,000 and the fixed costs £40,000. How many articles must this be in order to break even?
   If this number is only half of capacity, what profit will be made when production is at full capacity?

(5) If the total costs C to a company are related to revenue R by the equation

$$C = 0.35R + £22{,}750$$

find the total costs on revenue of £40,000 and the profit or loss on revenue of £30,000.
   At what revenue level is the break-even point?

2 / Linear Equations and Inequalities

SYSTEMS OF LINEAR EQUATIONS

In Chapter 1 we saw that the point where two lines intersect on a graph is the point where they are both satisfied by the same values of x and y. The graph is a picture of the equations. Two lines will intersect at a point unless they are parallel.

Suppose that you go into a cafe one day with some friends and buy for the group 5 cups of coffee and 2 cups of tea. The bill comes to 100p. How much is a cup of coffee? All that you know is that

$$100 = 5c + 2t \qquad (1)$$

where c is the price of a coffee and t the price of a tea. This equation clearly has more than one solution. Just two possibilities are: $c = 4$p with $t = 40$p; and $c = 8$p with $t = 30$p. Both are unlikely. The next day you go in again, and the person whose turn it is to buy is charged 88p for 3 coffees and 4 teas. Thus, we have

$$88 = 3c + 4t \qquad (2)$$

Assuming that no price changes overnight, we now have two linear equations in two unknowns (variables). These can be drawn as lines and the intersection of them used to give the prices. Alternatively, these *simultaneous equations* can be solved algebraically to find c and t.

We eliminate one variable, say t, by making it have the same coefficient in both equations. Multiplying (1) by 2 gives

$$200 = 10c + 4t \qquad (1a)$$

Subtracting (2) from (1a) gives

$$112 = 7c$$

$$c = 16 \text{ (i.e. 16p)}$$

Substituting $c = 16$ into (1) or (2), say (2), gives

$$88 = 3(16) + 4t$$
$$88 = 48 + 4t$$
$$40 = 4t$$
$$t = 10 \text{ (i.e. 10p)}$$

Substituting in (1) will check the answers.

A situation like this can arise in a firm making two products, say basic type and super type, which are processed in two departments, say P and Q. If they respectively take 5 minutes and 10 minutes in department P and 10 minutes and 12 minutes in department Q, with a total time available of 6 hours in P and 8 hours in Q, how many of each type must be produced to use the departments fully?

First, it is often useful to summarise the information in a table (see Table 2.1). Thus, if 40 basics and 20 supers are made,

$$5(40) + 10(20) = 400 \text{ minutes}$$

Table 2.1

Product	Minutes in each department	
	P	Q
	Required	
Basic	5	10
Super	10	12
	Total available	
	360	480

are needed in department P. If b of the basic and s of the super are made, department P has 360 minutes to spend on $5b$ minutes for the basics and $10s$ minutes for the supers. Hence, if department P is fully used,

$$360 = 5b + 10s \qquad (3)$$

Similarly, if department Q is fully used,

$$480 = 10b + 12s \qquad (4)$$

If (3) is multiplied by 2,

$$720 = 10b + 20s \qquad (3a)$$

Subtracting (4) from (3a) gives

$$240 = 8s$$
$$s = 30$$

Substituting back in (3) gives

$$360 = 5b + 10(30)$$
$$360 = 5b + 300$$
$$60 = 5b$$
$$b = 12$$

We shall generalise this idea later in the chapter on linear programming (Chapter 3). In practice, an availability of time or some other commodity is not always fully used, and we need to discuss this problem now.

INEQUALITIES

The room in which you are sitting now may have a maximum capacity of, say, 20 people. It is probably not full; and if x is the number of people in it now, perhaps $x = 8$. Suppose that X is the number of people that it could hold, where X is any number between 0 and 20. We write $X \leq 20$ (i.e. X is less than or equal to 20) and understand it to be a whole number (called an *integer*) and to be not negative.

A similar statement describing the fact that the height h of policemen cannot be below a minimum of 5 ft 8 in. would be $h \geq 68$ (h is greater than or equal to 68).

If the inequality strictly does not include the possibility of equalling the critical value, we would say, for example, $x > 4$ (i.e. x is a number bigger than 4) or perhaps $y < 7$ (i.e. y is a number less than 7).

Let us extend this idea to the time that is available in department Q in our recent example, which fully used is given by equation (4):

$$480 = 10b + 12s \quad \text{or} \quad 10b + 12s = 480$$

What if this time, which is clearly a maximum available, is not all used? Then

$$10b + 12s \leq 480$$

We must be even more careful when manipulating inequalities than when manipulating equations, but we can add to or subtract from each side and multiply or divide by positive numbers without reversing the inequality. Therefore, simplifying (dividing by 2), we get

$$5b + 6s \leq 240$$

This is no longer just a line, but what is it? It is the set of values of b and s (i.e. points) — such as $(10,20)$ and $(15,10)$ — that satisfy it. If we draw the line

$$5b + 6s = 240$$

we can see this more clearly (see Figure 2.1).

Incidentally, the easiest way to draw a line is often to say

$$\text{when } b = 0, \quad s = \frac{240}{6} = 40$$

$$\text{when } s = 0, \quad b = \frac{240}{5} = 48$$

by covering up each in turn with your finger on the equation. Either b or s can be the y-axis, as there is no clearly-defined dependence.

The difference between the equation and the inequality should now be clearer. It is the difference between the set of points *on* the line and the set of points, including $(10,20)$ and $(15,10)$, that lie *on it or below it*, which is called a *region*. It is like the difference between a county boundary and the area inside the county, but without the boundary you could not define the county. Note that, if we had

$$5b + 6s < 240 \quad (\text{not} \leqslant)$$

we would have the area below the line without the line too.

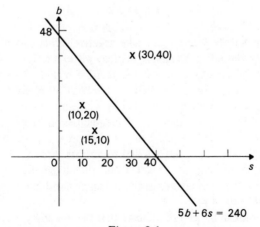

Figure 2.1

A quick check of another point that is likely to be outside the region, such as (30,40), shows that, when we substitute in the inequality, we get

l.h.s. = 5 (40) + 6 (30) = 380

That is, the left hand side (l.h.s.) is not less than the right hand side (r.h.s.) of 240, so this point does not satisfy the inequality. Substituting the origin (i.e. putting $s = 0$ and $b = 0$ in turn) is another quick check.

In order to show this region (the picture of the inequality) I prefer to shade out what is *not* satisfied. You will see why later. In Figure 2.2, what is left unshaded is where the inequality is satisfied (possibly including the

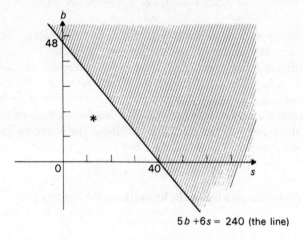

$5b + 6s = 240$ (the line)

Figure 2.2

boundary). If neither b nor s can take negative values (often the case in practice), only the positive triangle marked with an asterisk will be satisfied. This area is sometimes called the *feasible region*, since values of b and s within it are possible. Strictly, it contains infinitely many points, but again, in practice we may be limited by whole number possibilities.

Look carefully at the examples of inequalities in Figure 2.3, 2.4 and 2.5. Note in the first (Figure 2.3) that $x = 3$ and $y = 7$ are special simple cases of straight lines that are parallel to an axis. For $x = 3$, x is always 3, but y can be anthing, so we get a line that is parallel to the y axis. Similarly, for $y = 7$, y is always 7, but x can be anything, so we get a line that is parallel to the x axis.

A broken line can be used to indicate that the boundary is not included, but confusion is avoided by assuming this to be known, as in Figure 2.3.

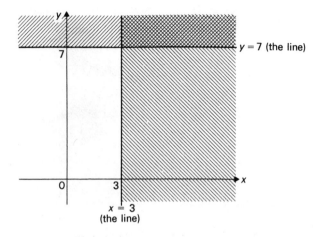

Figure 2.3 $x < 3$ and $y < 7$

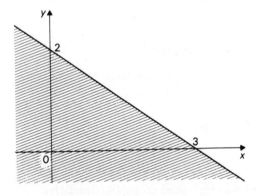

Figure 2.4 $2x + 3y \geqslant 6$

To draw the line in Figure 2.4, first say

$$\text{if } x = 0, \quad y = 2$$
$$\text{if } y = 0, \quad x = 3$$

Plot (0,2) and (3,0), and join.

Note that putting more than one inequality together usually further restricts the possibilities, as shown in Figure 2.5. Sometimes a constraint will be redundant because it will be included in another. For example, $y \geqslant 5$ and $y \geqslant 8$ can only happen together if $y \geqslant 8$; thus $y \geqslant 5$ is redundant.

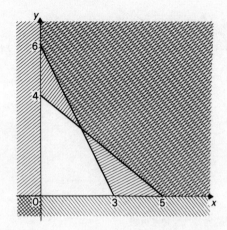

Figure 2.5 $y + 2x \leqslant 6$ and $5y + 4x \leqslant 20$
and $x \geqslant 0$ and $y \geqslant 0$
(both variables non-negative)

EXERCISES

(1) Solve the simultaneous equations

$$a + 2b = 5$$
$$3a - b = 1$$

(2) Solve simultaneously

$$4x + 3y = 11$$
$$6x + y = 6$$

(3) Show the following regions on graphs (separately):

(a) $y \geqslant 4$.
(b) $1 < x < 3$.
(c) $x + 2y < 0$.
(d) $3x + 2y \leqslant 18$.

(4) Use a graph to solve:

(a) $2x - y = 5$
$x + 3y = 13$

(b) $x + y > 6$
$2x + y > 8$

3 Linear Programming

Making use of the previous chapter, we shall now consider a technique often included under the subject umbrella of *Operational Research*. OR, as it is commonly called, is a collection of many applied techniques, where some mathematical idea or concept is used to solve a real problem. Often the problem has given rise to the new concept, in what is a fairly young branch of mathematics developed from studies of military problems during the Second World War.

The crucial factor in OR is that the problem is *analysed* mathematically to see what is really happening before a solution is sought. It is amazing how many problems are not what they appeared to be before the analysis and objective rationality got to work. How often have you heard people arguing about different things or about an irrelevant factor in their subject? There is an awful amount of padding, waffle and sheer entrenched nonsense to be stripped away from both business situations and social or personal situations.

Let logic be your tool or means, but do not forget that it is not necessarily an end in itself; your objectives or ends may be decided by aesthetic criteria.

We have already developed, with lines and inequalities, all the background that we need for *Linear Programming*, commonly known as LP. It remains only to use them in two full worked examples, after outlining situations where LP can be used as a model and the assumptions that we make.

We are generally concerned to *optimise* a quantity; that is, we want to get the best result from it according to some criterion, such as maximising profit or minimising cost. This optimisation has to take place subject to some *constraints* or limiting factors on our activity, such as the man-hours, cash or storage space that is available, or the minimum production level required by a contract.

As the name suggests, we assume that only linear functions, equations and inequalities are involved; for example, rates of production and prices of raw materials and of labour are assumed to be constant. We shall only concern ourselves at present with two variables, and our examples will therefore be two-dimensional and capable of graphical solution.

Example 1

A furniture company manufactures tables and chairs, which are processed through three departments. Each table takes 6 minutes in the Assembly Department, 10 minutes in the Finishing Department and 12 minutes in the Packing Department. The times for each chair in the three departments are respectively 12, 12 and 9 minutes. The total time that is available in a day is 12 hours in Finishing and Packing but only 10 hours in Assembly. There is a contract requiring 10 chairs to be produced every day, and the profits are £10 per table and £15 per chair.

We shall formulate this situation mathematically as a linear programming problem and find the product combination that gives the maximum profit.

We (always as a good rule) begin by letting our unknown combination of tables and chairs be x tables and y chairs. It is unimportant which we choose to be x or y.

Sometimes the constraints information is given in a table; but if not, it is helpful to express it as a summary as in Table 3.1.

Table 3.1

Department	Table	Chair	Total available
	Processing time (minutes)		
Assembly	6	12	600
Finishing	10	12	720
Packing	12	9	720
	Profit each (£)		
	10	15	

The number of chairs produced per day must be at least 10; that is,

$$\text{contract constraint} \quad y \geqslant 10$$

If the company is producing x tables taking 6 minutes each to process, that is $6x$ minutes taken up on tables. There are also $12y$ minutes taken to process the y chairs at 12 minutes each, so the total time needed in Assembly is $6x + 12y$. This cannot exceed the total of 600 minutes that are available (a 'less than or equal to' inequality); that is,

$$\text{assembly constraint} \quad 6x + 12y \leqslant 600$$

Similarly, we get

$$\text{finishing constraint} \quad 10x + 12y \leqslant 720$$
$$\text{packing constraint} \quad 12x + 9y \leqslant 720$$

We must not forget that $y \geqslant 10$.

The thing that we are attempting to optimise is called the *objective* function, which in this case is profit P, given by (in pounds)

$$P = 10x + 15y$$

We need to draw the inequalities and use our method of finding the boundaries (i.e. lines) by considering their equations and letting x and y be zero in turn to plot the lines. This method also has the important advantage of giving us a good guide to the scale to take on our axes. A little experience makes this quite easy. Thus,

$$\left.\begin{array}{l} 6x + 12y = 600 \\ \text{when } x = 0, \ y = 50 \\ \text{when } y = 0, \ x = 100 \end{array}\right\} \text{AA assembly}$$

$$\left.\begin{array}{l} 10x + 12y = 720 \\ \text{when } x = 0, \ y = 60 \\ \text{when } y = 0, \ x = 72 \end{array}\right\} \text{FF finishing}$$

$$\left.\begin{array}{l} 12x + 9y = 720 \\ \text{when } x = 0, \ y = 80 \\ \text{when } y = 0, \ x = 60 \end{array}\right\} \text{PP packing}$$

We need the x scale to include 100, 72 and 60, so we take 100 as convenient.

We plot the lines in Figure 3.1. Then, since the inequalities are 'less than or equal to', we shade out the region *above* each line, leaving clear the region that must therefore contain the values of x and y (the points) that satisfy *all* the inequalities at once. This is the *feasible region*. Any point (x, y) inside the feasible region is a combination of so many tables and chairs that can be produced. But which one gives the maximum profit?

Let us have a closer look at

$$P = 10x + 15y$$

This is actually not just a line but a *family of lines*, because P will vary depending on the values of x and y. If we rewrite it to get the form of a line

$$y = mx + c$$

which shows intercept c and gradient m as before, we get

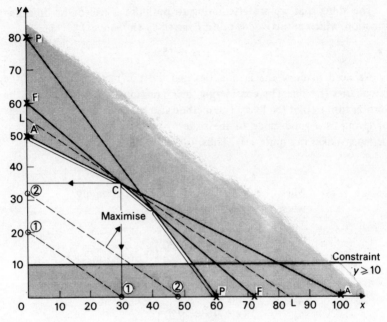

Figure 3.1

$$10x + 15y = P$$
$$15y = -10x + P$$
$$y = -\frac{2}{3}x + \frac{1}{15}P$$

Comparing
$$y = mx + c$$

We can see that the gradient m is always constant at $-\frac{2}{3}$ but that the intercept c can take any value $\frac{1}{15}P$. We therefore have infinitely many lines with the same gradient but different intercepts, which means that they are parallel. By allowing P to take two suitable values of 300 and 480, we can show two member lines of this family (dotted in Figure 3.1).

member ① $\quad 10x + 15y = 300$
member ② $\quad 10x + 15y = 480$

Looking at the form
$$y = -\frac{2}{3}x + \frac{1}{15}P$$

shows that higher values of P (which is what we are after) are given by larger intercepts $\frac{1}{15}P$, so we maximise by pushing the parallel member lines out further and further until we get the member shown as LL, which cuts as high an intercept as possible while *still* lying just in the feasible region at the corner C (called a *vertex*) as it 'leaves'. At C we therefore have the combination of x and y that is feasible *and* gives the greatest possible P. We can either read it from the graph or find it exactly by solving for the intersection of AA and FF by simultaneous equations.

Note that in practice the way to push parallel lines out is with a set-square and ruler, as shown in Figure 3.2.

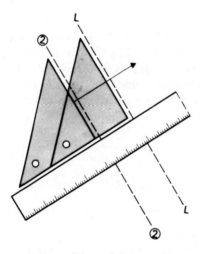

Figure 3.2

We get the (optimum) combination at C of $x = 30$ tables and $y = 35$ chairs. Substituting these back to find the maximum profit P_{max} gives

$$P_{max} = 10(30) + 15(35)$$
$$= 300 + 525$$
$$= 825 \quad (\text{i.e. £825})$$

Note the under-use made of Packing.

You will probably have realised that we have also used the obvious constraints $x \geqslant 0$ and $y \geqslant 0$, since we cannot produce negative quantities. The contract constraint happened to play no part in our answer but this cannot be taken for granted.

The optimum combination will always lie at one or other of the corners (called vertices) of the feasible region; and if the balance (i.e. ratio) of profits between the two products were altered, we would have to draw a different objective function, which might 'leave' the feasible region at a different vertex. Sometimes the profit balance is very sensitive to producing different answers.

A special case is when the objective function is actually parallel to one edge of the feasible region boundary. It then leaves the feasible region along all the points of that edge, and *any one of them* is an equally good optimum.

We shall now do another example to solve a minimisation problem.

Example 2

A certain diet for a week has to contain a minimum amount of three nutrients: 10 g of nutrient A, 12 g of nutrient B and 8 g of nutrient C. The nutrients are to be provided by two foodstuffs X and Y in quantities (grams per unit) shown in Table 3.2. The price per unit of each foodstuff is the same. Find the cheapest way of providing the diet.

Table 3.2

Foodstuff	Nutrient provided (g/unit)		
	A	B	C
X	5	4	3
Y	2	3	5

If we buy x units of foodstuff X and y units of foodstuff Y, clearly $x \geqslant 0$ and $y \geqslant 0$ in the usual way, which means that we are in the all-positive (i.e. first) quadrant of the Cartesian diagram (see Figure 3.3).

Our constraints are all that *at least* a certain amount of each nutrient is required. Hence, for nutrient A, 10 g must be exceeded by the $5x$ g from x units of foodstuff X, together with the $2y$ g from y units of foodstuff Y. Therefore,

$$\text{nutrient A constraint} \quad 10 \leqslant 5x + 2y$$

Similarly, we get

$$\text{nutrient B constraint} \quad 12 \leqslant 4x + 3y$$

$$\text{nutrient C constraint} \quad 8 \leqslant 3x + 5y$$

Our objective function is

$$S = x + y$$

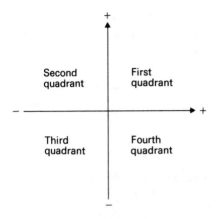

Figure 3.3

This is because the cost is given as unknown but the same for each foodstuff, say n pence/unit, giving as the total cost s (in pence)

$$s = nx + ny$$
$$= n(x+y)$$
$$\frac{s}{n} = x+y$$
$$y = -x + \frac{s}{n}$$

Notice that the only really important thing here is the gradient of -1. The units (pence or whatever) do not matter to the gradient; they merely change the intercept. As we saw in Example 1, it is the gradient of the line family that determines the optimum vertex. If we let

$$\frac{s}{n} = S$$

this simply changes the units, giving

$$y = -x + S$$
$$S = x + y$$

We plot the constraint boundaries in the usual way:

$$10 = 5x + 2y$$
when $x = 0$, $y = 5$
when $y = 0$, $x = 2$ } AA

$$12 = 4x + 3y$$
when $x = 0$, $y = 4$
when $y = 0$, $x = 3$ } $B_1 B_2$

$$8 = 3x + 5y$$
when $x = 0$, $y = \dfrac{8}{5}$
when $y = 0$, $x = \dfrac{8}{3}$ } CC

For the axis scales, y must include 5, 4 and $\frac{8}{5}$, and x must include 2, 3 and $\frac{8}{3}$. Thus, we take scales up to $y = 6$ and $x = 5$ for convenience, plot the lines and shade out areas *below* the lines to obtain the feasible region 'more than or equal to', where all the inequalities are satisfied together (see Figure 3.4). Notice that CC is made redundant by $B_1 B_2$.

To plot a typical member of the family of lines

$$S = x + y$$

we can either let S be a suitable number like 5, giving

$$5 = x + y$$

and points (0,5) and (5,0) to join up, or equivalently notice that any line with gradient -1 will do. Remember that the minus sign means simply a right-to-left slope.

We move the line in to find the corner that is still (just) feasible while having the lowest possible intercept, thus minimising S. This happens at B_2 giving the optimum combination (3,0) or $x = 3$ units and $y = 0$ units.

This result may surprise you, but it has been worked deliberately to illustrate that the rational optimum may be to buy only one foodstuff (i.e. X) and not to buy any of the other (i.e. Y). You may be able to think of industrial examples where a range of manufactured articles includes ones that should be discontinued. Buying 3 units of X along provides 15g of A, 12g of B and 9g of C, all of which quantities are sufficient. If the prices of X and Y change relative to one another, however, this may, of course, *not* be the optimum solution.

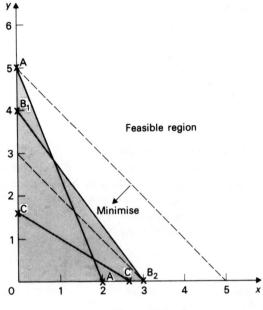

Figure 3.4

The redundancy of the CC constraint is due to the fact that the nutrient C requirement is met by the amounts of X and Y purchased to meet the A and B requirements. If the C requirement went up, say to 11 g, or the amount of C provided by one foodstuff went down, say to 2 g/unit for X, the constraint would possibly not be redundant.

The examples that we have taken may seem simple as models, but the technique of LP is a powerful one and can be extended (non-graphically) to more than two variables. The general model can also be extended to non-linear situations, giving rise to such things as *quadratic programming*.

One final point about the numbers that emerge in the answer for the optimum combination of x and y (either maximising or minimising): we obtained whole numbers in both cases, but what if they turn out not to be whole numbers?

For discrete items like tables and chairs we may accept such an answer if production is essentially continuous and odd parts are thought of as being made up into whole ones over a longer period. If whole numbers are strictly needed, we have to investigate the whole number points that are close to our answer to see which gives the greatest profit or lowest cost; for example, if we had our optimum point at a vertex like $(8\frac{1}{3}, 6\frac{1}{2})$, we would investigate the value of the objective function at points like (8,6) or (9,7), depending on context.

If the variable is continuous, there is no problem with an answer that is not a whole number; for example, a foodstuff like X could in theory be bought in quantities like $2\frac{2}{3}$ units.

EXERCISES

(1) A product is produced in two qualities, Super and Deluxe, which require amounts of the raw materials A and B as shown in the table (in suitable units). The daily availability of A is 24 units and of B 20 units. If the profits are £9 on each Super product and £8 on each Deluxe, find the daily combination that maximises profit, for both continuous and discrete values of product.

Raw material	requirement (units) Super	Deluxe
A	2	3
B	5	2

(2) A car manufacturer has two models of car, the Happy and the Midi, which pass during manufacture through three areas or 'shops': the Body Shop (for assembly), the Spray Shop (for painting) and the Service Shop (for checking). Each Happy takes 4 man-hours, 1 man-hour and 5 man-hours respectively to pass through, and each Midi 3 man-hours, 3 man-hours and 2 man-hours respectively. The weekly availability in each shop is 1,200 man-hours in Body, 1,020 in Spray and 1,200 in Service. Because the Happy is in a relatively more competitive market sector than the Midi, the profits are £100 on each Midi and £50 on each Happy. If the company assumes that any production can be sold, what production policy will maximise profits?

If the company assumes that at most 300 Midis can be sold weekly, will the optimum policy be different?

(3) The owner of an electrical shop is to stock up with televisions, radios and cassette players, but he is limited by the cash that is available to pay for the order and the storage space in the shop. Each item has its own cost, volume and profit as shown in the table, which also shows the cash and space limits. The shopowner also decides to stock at least

40 radios. Formulate this situation as a linear programming problem in three variables.

If he does not stock cassette players after all, what will be his 'best' order?

Item	Cost (£)	Volume (ft³)	Profit (£)
	Required		
Television	40	3	12
Radio	15	1.5	4.5
Cassette player	20	1	8
	Limits on availability		
	2,000	180	

(4) Mr and Mrs J. Spratt plan their meat purchases for a month by establishing the following criteria. An adult needs at least 18 g of vitamin A and 30 g of vitamin B every month for good health. The only meats under consideration to provide these vitamins are lamb at 90 pence/lb, which contains 5 g/lb of vitamin A and 3 g/lb of vitamin B, and pork at 72 pence/lb, which contains 3 g/lb of vitamin A and 6 g/lb of vitamin B. Formulate their predicament as a linear programming problem, and find the cost of the cheapest solution.

If the prices of lamb and pork changed so that lamb cost twice as much as pork (with everything else unchanged), what would be the new optimum combination?

(5) A nurse needs to administer a daily liquid diet to a patient so that he receives a minimum of 3 units of vitamin P and 4 units of vitamin U. He needs a total intake of at least 3 pints of liquid, which can be lime juice and/or blackcurrant cordial. The vitamin content of each liquid and their costs are given in the table. Find the optimum way of providing the diet and its cost.

Liquid	Vitamin content (units/pint)		Cost (pence/pint)
	P	U	
Lime juice	0.75	2	5
Blackcurrant cordial	2	1	8

(6) A mining company has a contract to supply not less than the annual tonnages of certain mineral ores shown in the table, which also includes the weekly cost of working the company's two mines and the weekly tonnages of the mineral ores that each mine produces. Find the number of weeks per year that each mine should be worked in order to minimise the company's costs. (Assume that the mines have an annual two-week maintenance shut-down.)

If the costs were reversed (i.e. £120 per week for the Lucky Strike and £150 per week for the Last Chance), what would be the optimum operating policy?

Mine	Mineral ore (tons)			Cost (£)
	L	M	N	
	Weekly production			
Lucky Strike	1	2	4	150
Last Chance	2	2	1	120
	Annual contract			
	120	150	100	

4 Differential Calculus

FUNCTIONS

We have already met the idea of a linear relation and seen some examples of it. In general we have seen that

$$y = mx + c$$

is the equation of a straight line.

We can expand the concept to define a *function*. For any value of x we get a unique value of y. I said before that x is the independent variable (in practice because it appears first); y depends on x or is said to be a function of x. We write

$$y = f(x)$$

for the values of x under consideration. This simply means that y is related to x by some rule

$$\text{e.g.} \quad y = 2x + 3$$

so that every value of x is associated with one value of y.

We have already constructed examples, such as costs related to number produced. Another important example in Economics considers demand D to be a function of price p, or

$$D = f(p)$$

In simple price theory we use a linear model

$$D = ap + b$$

where a and b are constants; for example,

$$D = 100 - 5p$$

The relationship may, of course, be more complicated, involving higher powers of p or reciprocals. The crucial point is that, given any price p, there is a demand D. Does it look a realistic connection?

We shall see shortly the importance of a type of function known as a *polynomial* ('many powers'). This simply means that many powers are possible. A polynomial can have any power appearing but does *not* contain other things that you may meet elsewhere (e.g. $\sin x$ or e^x). Examples of polynomials are

$$\text{quadratic function} \quad y = 2x^2 - 3x + 5$$
$$\text{cubic function} \quad y = x^3 + 4x^2 - 11$$

The *degree* of a polynomial is the highest power that appears.

The function notation $y = f(x)$ or $D = g(p)$ is very useful for evaluations. (Note that $g(p)$ indicates that D is some function or rule g of p that is different from f).

Suppose that demand D depends on price p (i.e. is a function of p) thus:

$$D = g(p)$$
$$\text{e.g.} \quad D = 10 - p$$

and we want to know the demand at prices 2, 5 and 10. Substituting these values in

we get
$$g(p) = 10 - p$$
$$D = g(2) = 8$$
$$D = g(5) = 5$$
$$D = g(10) = 0$$

Let us now further suppose that cost C is a function of (i.e. depends in some way on) demand D; that is,

$$C = f(D)$$
$$\text{e.g.} \quad C = 5 + D^2$$

which is a quadratic function.

Notice that, since C depends on D and D depends on p, then C is linked to, and depends on, p. This situation is called *function of a function*. For example, imagine yourself waiting for a friend under the clock at Victoria Station, while the friend is sheltering two streets away waiting for the rain to cease and join you; like it or not, you are yourself dependent on the rain.

We can evaluate C (i.e. evaluate $5 + D^2$) by substituting the value of D corresponding to a certain value of p. For example, if $p = 2$,

$$D = 10 - p$$
$$= 10 - 2 = 8 \quad \text{(as before)}$$
$$\therefore \quad C = 5 + D^2$$
$$= 5 + (8)^2 = 69$$

when $p = 2$. More generally, we can substitute for D into the C function:
$$C = 5 + D^2$$
$$= 5 + (10 - p)^2$$
$$= 5 + 100 - 20p + p^2$$
$$= 105 - 20p + p^2$$

The last equation is the general relation between C and p (another function).

DIFFERENTIATION

You will eventually use the calculus as a tool, by applying simple rules to certain situations, without thinking about it twice, but it is extremely important that you try to understand the concepts that we are about to discuss before developing the rules.

The idea of infinitely small divisions of time or space and the concept of a limit have puzzled men at least since Zeno first expounded his paradoxes. Just one of them is as follows. An arrow fired from an archer to its target will first have to travel to the halfway point. Before this it will naturally have to reach the halfway point of halfway, and so on. Since this must be repeated any number of times, repeatedly bisecting a very small but still finite distance, it is clear that the arrow can never reach the target.

This is not an easy argument to grasp as being incorrect, because the idea of an infinite series of numbers adding up to one finite number is a difficult one.

Although the rigorous basis to the calculus was not developed until long after the seventeenth century, problems then causing concern about tangents to a curve and areas within a curve caused Newton and Leibnitz to produce the immensely powerful tool of the calculus.

We are fundamentally concerned with the rate at which one variable is changing with respect to another, instantaneously. We have already met the gradient of a straight line, which is a rate of change, albeit a constant one. In particular, a horizontal line (e.g. $y = 5$) has gradient zero, because it is flat and is not rising.

For a curve this rate of change is not so simple; it is changing at every point. The curve is obviously rising faster at Q than at P, for example, in Figure 4.1. In fact we would not be able to define the rate of change in the same way as for a straight line unless we said that *the rate of change (i.e. gradient) of a curve at a point is equal to the gradient of the tangent at that point.* We can then calculate the gradient at P in the usual way as

$$\frac{\text{Vertical distance risen}}{\text{Corresponding horizontal distance}} = \frac{v}{h}$$

Note that *any* such triangle on the tangent will give this ratio, as they are all similar triangles. Incidentally, by definition,

$$\frac{v}{h} = \tan \theta$$

where θ is the angle between the tangent and the x axis.

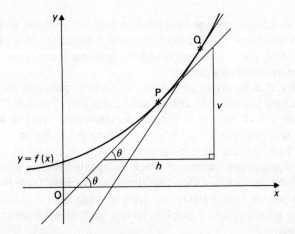

Figure 4.1

How do we connect this gradient to the function (i.e. the curve)?

Consider two points very close together on the curve. These are shown large merely to demonstrate them in Figure 4.2. Here, R and S are supposed to be so close together that, if R is the point (x, y) (i.e. OM = x and RM = y), S is such that its x value (called its *x coordinate*) ON has increased to $(x + h)$ by a very small amount h. Using our functional notation

Differential Calculus 33

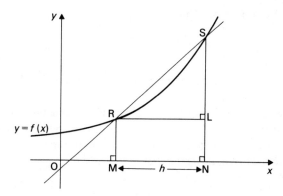

Figure 4.2

We get
$$y = f(x)$$
$$RM = f(x)$$
$$SN = f(x + h)$$

Now,
$$\text{Gradient of RS} = \frac{SL}{RL}$$
$$= \frac{SN - LN}{MN}$$

or, in functional notation,
$$\text{Gradient of RS} = \frac{f(x + h) - f(x)}{h}$$

Now, RS is not yet a tangent (actually it is a chord); but as S gets nearer to R (we write S → R, or S tends to R), in the limit the chord becomes a tangent, and the small increase h tends to 0 as the points merge. We write

$$\text{Gradient of tangent} = \lim_{h \to 0} \left\{ \frac{f(x + h) - f(x)}{h} \right\}$$

We have obtained this definition from first principles.

The usual notation used for this limit (i.e. the gradient of the tangent) is

$$\frac{dy}{dx} \quad \text{or} \quad \frac{d}{dx} f(x)$$

which are the same thing. Thus

$$\frac{dy}{dx} = \lim_{h \to 0} \left\{ \frac{f(x+h) - f(x)}{h} \right\} \quad \text{where } y = f(x)$$

There are other notations, principally $f'(x)$, $Df(x)$ and $\dot{x}$; but remember that they are all arbitrary.

Remember that dy/dx does not really represent one number divided by another but rather $dx/$a limit, a process, an instantaneous rate of change, an operation d/dx acting on y. This is most important. The operation is called the *derivative* or *differential* of y.

Consider

$$y = x^2 - x$$

between $x = -1$ and $x = 3$ (see Figure 4.3).

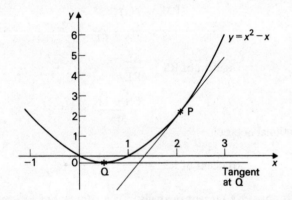

Figure 4.3

What is the gradient at $x = 2$?
We start with

$$y = f(x) = x^2 - x$$

Using the definition just derived gives

$$\frac{dy}{dx} = \frac{d}{dx} f(x) = \lim_{h \to 0} \left\{ \frac{f(x+h) - f(x)}{h} \right\}$$

$$= \lim_{h \to 0} \left\{ \frac{[(x+h)^2 - (x+h)] - (x^2 - x)}{h} \right\}$$

Notice particularly the substitution of $(x + h)$ in the function. Expanding and simplifying give

$$\frac{dy}{dx} = \lim_{h \to 0} \left\{ \frac{x^2 + 2xh + h^2 - x - h - x^2 + x}{h} \right\}$$

$$= \lim_{h \to 0} \left\{ \frac{2xh + h^2 - h}{h} \right\}$$

$$= \lim_{h \to 0} \{2x + h - 1\}$$

Letting h go to zero gives

$$\frac{dy}{dx} = 2x - 1$$

Therefore, at P, where $x = 2$, the gradient or derivative is given by

$$\left(\frac{dy}{dx}\right)_{x=2} = 2(2) - 1 = 3$$

You should be able to see by eye from Figure 4.3 that at P the gradient is indeed about $+3$ (rising left to right three up for every one along).

Although using the result

$$\frac{dy}{dx} = 2x - 1$$

for the derivative of

$$y = x^2 - x$$

gives it to us at any point on the curve just by substituting the value of x at that point, we shall never in practice find derivatives from first principles. There is *one simple rule* for any polynomial term:

if $\quad y = ax^n$

then $\quad \dfrac{dy}{dx} = anx^{n-1}$

where a and n are constants; for example,

if $\quad y = 3x^7$

then $\quad \dfrac{dy}{dx} = 21x^6$

Terms connected by + or − signs are simply done separately; for example

$$\text{if} \quad y = 4x^5 - 11x^4 + 2x^3$$

$$\text{then} \quad \frac{dy}{dx} = 20x^4 - 44x^3 + 6x^2$$

A particular case is when $n = 1$; for example,

$$\text{if} \quad y = 5x$$

$$\text{then} \quad \frac{dy}{dx} = 5(1)x^{1-1} = 5x^0 = 5$$

since anything raised to the power 0 is equal to 1. (Try to prove it using the indices multiplication and division rules of algebra.) This, of course, tallies completely with our previous knowledge that 5 was the gradient of the line $y = 5x$.

If $n = 0$, the rule still applies; for example,

$$\text{if} \quad y = 11 \quad (= 11x^0 \text{ since } x^0 = 1)$$

$$\text{then} \quad \frac{dy}{dx} = 11(0)x^{0-1} = 0$$

This also confirms that the gradient of a horizontal line is zero. So, if y equals any constant, $dy/dx = 0$, but this is really a·special case of the general rule.

We can check our example

$$y = x^2 - x$$

by the rule:

$$\frac{dy}{dx} = 2x^{2-1} - 1x^{1-1} = 2x^1 - 1x^0 = 2x - 1$$

If the power of x is negative or fractional the same rule still applies; for example,

$$\text{if} \quad y = \frac{2}{x^3} = 2x^{-3}$$

$$\text{then} \quad \frac{dy}{dx} = 2(-3)x^{-3-1} = -6x^{-4} = -\frac{6}{x^4}$$

and

if $y = -3x^{1/2}$

then $\dfrac{dy}{dx} = -3\left(\dfrac{1}{2}\right)x^{1/2-1} = -\dfrac{3x^{-1/2}}{2} = -\dfrac{3}{2x^{1/2}}$

A good example of an easily recognisable rate of change is velocity. A moving car at any moment changes its position with respect to time, so we can write

$$v = \dfrac{dx}{dt}$$

where v is velocity, x is distance and t is time. The difference between velocity and speed is that speed is a *scalar* and has size (magnitude) only whereas velocity is a *vector* and has both size and direction (e.g. a velocity of 40 km/hour due north).

Can this concept of rate of change be extended further? You have only to use the car example again to see that a car changes its velocity with respect to time by accelerating. So, we can write

$$a = \dfrac{du}{dt}$$

where a is acceleration. Substituting for v we get

$$a = \dfrac{d}{dt}\left(\dfrac{dx}{dt}\right)$$

So, acceleration is a *rate of change of a rate of change*. We call this a *second differential* or *second derivative* and write it, in this case, as

$$a = \dfrac{d^2x}{dt^2}$$

Although the practical meaning is not easy to see (as it is for velocity or acceleration), we can continue indefinitely, getting

third differential $\dfrac{d^3x}{dt^3}$

fourth differential $\dfrac{d^4x}{dt^4}$

and so on. To calculate these we just apply the rule repeatedly; for example,

$$\text{if} \quad y = x^2 - x$$

$$\text{then} \quad \frac{dy}{dx} = 2x - 1$$

$$\frac{d^2y}{dx^2} = 2$$

$$\frac{d^3y}{dx^3} = 0$$

An extremely useful idea now ready to be developed is that of *maxima and minima*. Referring back to Figure 4.3, i.e. to the function

$$y = x^2 - x$$

we notice that it has a minimum point; that is, the curve comes down, levels off and starts to rise again at θ. Here, $x = \frac{1}{2}$ and y (by substitution) is

$$(\tfrac{1}{2})^2 - (\tfrac{1}{2}) = -\tfrac{1}{4}$$

This value (i.e. $y_{\min} = -\tfrac{1}{4}$ at Q) is the *minimum value* of the function.

Can we calculate where it occurs in general, using our calculus?

You should see at once that what is special about a minimum point (and, equally, about a maximum point) is that, because the curve is turning at that point, the tangent to the curve there is horizontal and therefore has a gradient of zero. Thus, if we calculate the gradient dy/dx and *put* it equal to zero, we shall know the value of x that makes it zero (i.e. where it happens). Hence,

$$\text{if} \quad y = x^2 - x$$

$$\text{then} \quad \frac{dy}{dx} = 2x - 1 = \text{gradient of curve}$$

At turning points

$$2x - 1 = 0$$
$$2x = 1$$
$$x = \tfrac{1}{2}$$

Actually, to find the minimum value of y we must substitute back, as before.

How can we tell in general whether we have a max. or a min., as they are commonly called, without drawing the curve?

In Figure 4.4 you will see two arbitrary curves with tangents drawn around the turning points. The gradients of the tangents near the max. are changing from + (sloping left to right) through 0 (horizontal at the max.) to − (sloping right to left) and are therefore decreasing. Thus, the rate of change of the curve (i.e. the gradient) is itself undergoing a negative rate of change. Since we call this rate of change of a rate of change the second derivative d^2y/dx^2, a negative value for it will indicate a max.

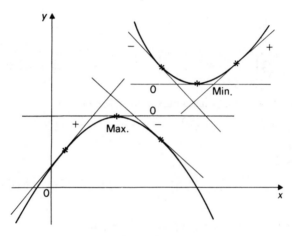

Figure 4.4

Exactly the same reasoning on the other curve tells us that a positive value for d^2y/dx^2 indicates a min. For

$$y = x^2 - x$$

we found $d^2y/dx^2 = 2$, which is positive (its actual numerical value is not important) and confirms that we should have a min.

Other possibilities (e.g. *inflexions*) you may like to inquire about, but they do not concern us here.

Example 1

Find the maximum and minimum values of

$$y = x^3 - x^2 - x + 4$$

General gradient is given by

$$\frac{dy}{dx} = 3x^2 - 2x - 1$$

At turning points the gradient will be zero; that is,

$$3x^2 - 2x - 1 = 0$$
$$(3x + 1)(x - 1) = 0$$
$$x = -\tfrac{1}{3} \text{ or } 1$$

Now,
$$\frac{d^2y}{dx^2} = 6x - 2$$

When $x = -\tfrac{1}{3}$,

$$\left(\frac{d^2y}{dx^2}\right)_{x=-1/3} = 6\left(-\frac{1}{3}\right) - 2 = -4 \quad \text{which is negative}$$

Therefore, at $x = -\tfrac{1}{3}$, y has a maximum value and

$$y_{\max} = (-\tfrac{1}{3})^3 - (-\tfrac{1}{3})^2 - (-\tfrac{1}{3}) + 4 = -\tfrac{1}{27} - \tfrac{1}{9} + \tfrac{1}{3} + 4 = 4\tfrac{5}{27}$$

When $x = 1$,

$$\left(\frac{d^2y}{dx^2}\right)_{x=1} = 6(1) - 2 = 4 \quad \text{which is positive}$$

Therefore, at $x = 1$, y has a minimum value and

$$y_{\min} = (1)^3 - (1)^2 - (1) + 4 = 3$$

This knowledge considerably helps the job of drawing the curve.

ECONOMICS APPLICATIONS

Although various applications have been considered already in passing, one or two other new ideas will now be aired, as we bring all the concepts and techniques developed thus far into a full worked example.

Example 2
The manufacturer of Northern Cookies (a delicacy that has not yet spread south of Watford Gap) believes that the total cost C of making q thousand cookies is given by

$$C = 150 + \tfrac{3}{2}q$$

(Note: All measurements of money are in suitable units.) The demand curve relating price p and quantity made is given by

$$p = 50 - q$$

Thus, a larger quantity produced will reduce the individual price.

Let us suppose that the manufacturer wishes to maximise revenue R rather than profit P. (The criteria for deciding the objective to be optimised do not always lead to the obvious, as a consideration of current affairs may show you.)

$$\text{Total revenue} = \text{Quantity produced} \times \text{Price each}$$

$$R = qp$$

or, if we make R a function of q alone,

$$R = q(50 - q)$$
$$= 50q - q^2$$

Hence,

$$\frac{dR}{dq} = 50 - 2q$$

At turning points $dR/dq = 0$; that is,

$$50 - 2q = 0$$
$$q = 25$$

Now,

$$\frac{d^2R}{dq^2} = -2 \quad \text{which is negative}$$

Therefore, maximum revenue $R_{\max}$ is when 25,000 are produced. Substituting back gives

$$R_{\max} = 25(50 - 25) = 625$$

Let us now consider maximising the profit.

$$\text{Total profit} = \text{Total revenue} - \text{Total costs}$$

$$P = R - C$$
$$= (50q - q^2) - (150 + \tfrac{3}{2}q)$$
$$= 50q - q^2 - 150 - \tfrac{3}{2}q$$
$$= -q^2 + \tfrac{97}{2}q - 150$$

Differentiating, we get

$$\frac{dP}{dq} = -2q + \tfrac{97}{2}$$

Hence, for a maximum or minimum,

$$-2q + \tfrac{97}{2} = 0$$
$$q = \tfrac{97}{4}$$

This is quite feasible, as q is measured in thousands. Also,

$$\frac{d^2P}{dq^2} = -2 \quad \text{which is negative}$$

Thus, maximum profit $P_{\max}$ is when 24,250 are produced. Substituting back $q = \tfrac{97}{4}$ will give the value of $P_{\max}$.

An alternative but equivalent method uses the idea of *marginal cost MC* and *marginal revenue MR*. Marginal cost, which we met briefly before as a variable cost per item in a linear model, is the rate at which total cost TC is changing with respect to quantity — a sort of instantaneous 'cost of the next one'. Therefore,

$$MC = \frac{dTC}{dq}$$

Marginal revenue is defined in exactly the same way:

$$MR = \frac{dTR}{dq}$$

where TR is total revenue.

Now, if these two things are equal, the cost of the next item produced equals the revenue from this next item produced, so our profit position is not changing (i.e. is stationary) at that moment. This corresponds exactly to our maximum before, and so maximum profit is where $MC = MR$. Care must be taken as they are also equal for a minimum, however.

In our example

$$\frac{d}{dq}(150 + \tfrac{3}{2}q) = \frac{d}{dq}(50q - q^2)$$

$$\tfrac{3}{2} = 50 - 2q$$

and so on, giving the same answer as above.

We can use the same general method to optimise other things, for example, to minimise the average cost AC. Remember that

$$AC = \frac{TC}{q}$$

ELASTICITY

We have already met examples of demand functions where the quantity demanded is a function of the price; for example,

$$q = 52 - p - p^2$$

As the price goes up (or down), we would expect the quantity demanded to go down (or up); but the rate at which q is changing with respect to (w.r.t.) p, which in differential form is dq/dp, is not quite what the *elasticity* of q at a certain p measures. Change *relative* to the values that q and p have at a particular point is much more meaningful.

The general definition of the *elasticity of* $y = f(x)$ *at* x is the rate of proportional change in y per unit proportional change in x; that is

$$E_{yx} = \frac{x}{y} \cdot \frac{dy}{dx}$$

It may be easier for you to match the symbols to the written definition if you think of them as

$$\frac{dy/y}{dx/x}$$

but remember this is just to show the proportions and does not really represent four numbers at a point on a continuous curve; d/dx is an operator.

In our economics example above the *elasticity of demand* (quantity) at price p is η (eta), where

$$\eta = \frac{p}{q} \cdot \frac{dq}{dp}$$

Another way to think about η is as a measure of the response of market demand when the price is changed by a small amount. Our example gives

$$\frac{dq}{dp} = -1 - 2p$$

$$\eta = \frac{p}{q}(-1 - 2p)$$

$$= \frac{p}{q}(1 + 2p)$$

Knowing the particular values of p and q at a point, we can calculate a numerical value for η. Note that substitution for q from the original equation would give η as a function of p only.

The *elasticity of supply* (quantity) at price p is ϵ (epsilon), found in exactly the same way. If we have the supply function

$$q = 2(20 + 4p + p^2)$$

$$\therefore \quad \frac{dq}{dp} = 8 + 4p$$

$$\therefore \quad \epsilon = \frac{p}{q}(8 + 4p)$$

$$= 4\frac{p}{q}(2 + p)$$

A typical economics example would ask for the two elasticities when demand and supply balance (i.e. at **market equilibrium**), so let us calculate them here. If demand equals supply,

$$52 - p - p^2 = 40 + 8p + 2p^2$$
$$0 = -12 + 9p + 3p^2$$
$$p^2 + 3p - 4 = 0$$
$$(p - 1)(p + 4) = 0$$
$$p = 1$$

which is the price at equilibrium. When $p = 1$,

$$\text{Demand} = 52 - 1 - (1)^2 = 50$$
$$\text{Supply} = 40 + 8(1) + 2(1) = 50$$

Substituting these values for q gives

$$\eta = -\tfrac{1}{50}[1 + 2(1)] = -\tfrac{3}{50} \quad \text{for demand}$$
$$\epsilon = 4(\tfrac{1}{50})(2 + 1) = \tfrac{12}{50} \quad \text{for supply}$$

DIFFERENTIATION OF THE EXPONENTIAL FUNCTION e^x

You are used to seeing functions like

$$y = x^2$$

where the variable x is raised to a constant power; and there is no reason to be frightened by a function like

$$y = 2^x$$

where a constant is raised to a variable power. We can plot such a function by substituting values of x in the usual way, as Table 4.1 and Figure 4.5 show.

Table 4.1

x	-1	0	1	2	3	4	5	6	etc.
$y = x^2$	1	0	1	4	9	16	25	36	(for comparison)
$y = 2^x$	$\frac{1}{2}$	1	2	4	8	16	32	64	(see Figure 4.5)

Any graph of a constant raised to a variable power, i.e.

$$y = a^x$$

has a similar shape. They come up from zero through the point (0,1), since anything to the power 0 equals 1, and then climb rapidly to infinity. One in particular is very important: that when the constant has the particular value 2.7183 approx., and this is known as e. The function is therefore

$$y = e^x$$

Remember that e is just a constant. Most books of tables include an e^x table.

The function e^x can be defined as

$$e^x = 1 + \frac{x}{1!} + \frac{x^2}{2!} + \frac{x^3}{3!} + \frac{x^4}{4!} + \ldots$$

which is an infinite series involving **factorial terms** (explained in Chapter 13). Putting $x = 1$, we get

$$e^1 = 1 + \frac{1}{1!} + \frac{1}{2!} + \frac{1}{3!} + \frac{1}{4!} + \ldots$$

Taking relatively few terms, we can calculate e as 2.7183 approx.

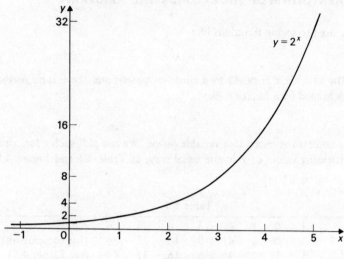

Figure 4.5

One of the very important things about e^x is that it differentiates to itself — the only function that does so. This means that its rate of change at any point is equal to its value at that point. Differentiating the e^x series term by term, we get

$$\frac{d}{dx}(e^x) = 0 + \frac{1}{1!} + \frac{2x}{2!} + \frac{3x^2}{3!} + \frac{4x^3}{4!} + \ldots$$

$$= 1 + x + \frac{x^2}{2!} + \frac{x^3}{3!} + \ldots$$

Cancelling into each factorial gives the same series again.

$$\frac{d}{dx}(e^x) = e^x$$

A constant present in the function gives the similar result; for example,

$$\frac{d}{dx}(e^{kx}) = ke^{kx}$$

An economics example would be a supply function defined by

$$q = e^{3p}$$

Hence,

$$\frac{dq}{dp} = 3e^{3p}$$

and elasticity of supply is

$$\epsilon = \frac{p}{q} \cdot \frac{dq}{dp}$$

$$= \frac{p}{e^{3p}} \cdot 3e^{3p}$$

$$= 3p$$

EXERCISES

(1) Sketch the function

$$y = 3x^2 - 12x + 4$$

by substituting suitable values of x.

(2) Calculate $f(1), f(2)$ and $f(x+h)$ if

$$f(x) = 2x + 5x^2$$

(3) Use the first-principles definition to find, at $x = 2$, the instantaneous rate of change of the function

$$y = x^3 + x - 7$$

(4) Differentiate with respect to x:
   (a) $y = 17 + 4x + 2x^2$.
   (b) $y = 1{,}000{,}000 - 7x^5$.
   (c) $y = 5x^3 - x + 3$.

(5) Find the gradient of the tangents to the curve

$$y = 2x^2 - 4x + 5$$

at the points where $x = 1$ and $x = 2$.

(6) Differentiate with respect to p

$$D = \frac{a}{p^2} + b$$

(7) Differentiate with respect to x

$$y = \frac{3}{x^2} - \frac{4}{x} + 7x^{1/3}$$

(8) Find any max. and min. values of the function

$$y = x^2 - 4x - 2$$

Sketch the curve.

(9) A manufacturer of cardboard boxes wants to make open boxes out of pieces of cardboard 30 in. square, by cutting squares out of each corner and folding up the sides. Show that, if x is the side of these squares (cut out), the volume of the boxes is given by

$$V = 4x^3 - 120x^2 + 900x$$

What size squares should be cut out of the corners of the pieces of cardboard so as to maximise the volume of these open boxes?

(10) An importer claims that, if the government charges a tax of £t on an ounce of a certain kind of perfume, imported from an Asian country, he will be able to sell $(45{,}000 - 300t)$ oz of the perfume. Assuming this claim to be correct, what tax should the government charge so as to maximise its total revenue on this perfume?

(11) A shopkeeper buys ornaments at £1 each. He believes that, if he sells them for £x each, he will be able to sell $10(12-x)$ ornaments. At what selling price will he maximise his total profit?

(12) In an industry that has total costs given by

$$C = \frac{q^2}{100} + \frac{q}{10} + 40$$

and demand curve

$$p = 20 - \frac{q}{25}$$

find the value of q that maximises total profits.

(13) If the demand function for a firm's product is given by

$$p = 302 - 0.01q$$

and the total cost function is given by

$$C = 0.01q^2 + 2q + 100$$

using calculus, find an expression for:
(a) marginal revenue.
(b) marginal cost.
Find the level of output at which profits would be maximised.

(14) (a) If

$$\text{Total revenue} = 100q - 2q^2$$

$$\text{Total costs} = 400 + 40q$$

what is the value of q that maximises profit?
(b) A cost function is given by

$$C = 0.2q^2 + 4q + 300$$

where q denotes the output. Sketch this function.
What is the average cost function?
For what output is the average cost of production minimised?
What is the marginal cost function?
Sketch this and the average cost function on the same graph.
Where do the two graphs intersect?
What does this point indicate?

5 / Partial Differential Calculus

DIFFERENTIATION REVISION

We have already met the idea of a function of one variable and its rate of change, called the derivative or differential. Given

$$y = f(x) \quad \text{(a line or curve)}$$
$$\text{e.g. } y = x^2 - 4x + 3 \quad \text{(a parabola)}$$

rate of change of y with respect to x is dy/dx; in this example, using the rule of page 35, we get

$$\frac{dy}{dx} = 2x - 4$$

This rate of change d/dx, which is an *operator* (i.e. a process performed on y), measures the gradient of the tangent to the curve, since this shows how the curve is changing.

To get a picture of this and revise some ideas, let us plot the curve, by substituting different values of x (see Figure 5.1):

$$\text{when } x = 0, \quad y = f(0) = (0)^2 - 4(0) + 3 = 3$$
$$x = 1, \quad y = f(1) = (1)^2 - 4(1) + 3 = 0$$
$$x = 2, \quad y = f(2) = (2)^2 - 4(2) + 3 = -1$$
$$x = 3, \quad y = f(3) = (3)^2 - 4(3) + 3 = 0$$
$$x = 4, \quad y = f(4) = (4)^2 - 4(4) + 3 = 3$$
$$\text{etc.}$$

The derivative is

$$\frac{dy}{dx} = 2x - 4$$

which at B has value

$$2(3) - 4 = 2$$

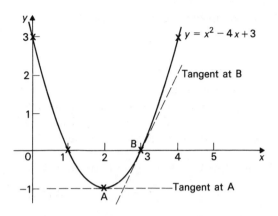

Figure 5.1

by substituting $x = 3$ at B. At A, where $x = 2$, the derivative has value

$$2(2) - 4 = 0$$

This is because the tangent at A is horizontal and must have a gradient of zero.

We have a minimum value (or turning point in general) at A. Remember, we differentiate dy/dx again to determine the nature of the turning point. Thus, using the same rule again, we get

$$\frac{d^2y}{dx^2} = 2$$

Since this is positive, we have a minimum. If it were negative, we would have a maximum.

PARTIAL DIFFERENTIATION

I have bothered to revise all this in part because the ideas that we need for partial differentiation are just an extension of these previous ideas, which is necessary because the world in which we live is three-dimensional, not the two-dimensional one of Figure 5.1.

We easily achieve this by extending the function of one variable

$$y = f(x)$$

to a function of two variables, usually expressed as

$$z = f(x, y)$$

e.g. $z = 2x^2y + xy - 1$

This represents a *surface* in three-dimensional space. Simply substituting values of x and y gives a value of z, which must describe a point. Continual substitution gives all the points that make up the surface. To plot the above function, for example (see Figure 5.2), we say

if $x = 0$ and $y = 0$, then $z = -1$ point A $(0, 0, -1)$

$x = 1$ $y = 2$ $z = 5$ B $(1, 2, 5)$

$x = 2$ $y = 1$ $z = 9$ C $(2, 1, 9)$

etc.

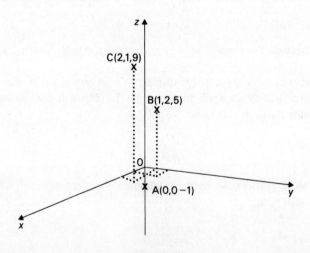

Figure 5.2

The shape of the surface that we would actually obtain is far too complicated to draw accurately here. A typical surface would look like the one in Figure 5.3.

The big difference in a function of two variables like

$$z = 2x^2y + xy - 1$$

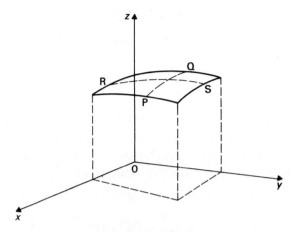

Figure 5.3

is that a change in *either x or y* causes a change in z; for example,

if $x = 1$ and $y = 1$, then $z = 2$

$x = 1 \quad y = 2, \quad z = 5$ (as before)

$x = 2 \quad y = 1 \quad z = 9$ (as before)

z has changed from 2 to 5 because y has changed from 1 to 2 while x has remained constant at 1. Similarly, z has changed from 2 to 9 because x has changed from 1 to 2 while y has remained constant at 1.

This means that the function z must have a rate of change with respect to both x and y. Figure 5.3 shows this pictorially, because it is possible to move over the surface in two well-defined mutually-perpendicular directions: parallel to the xz plane (i.e. along PQ) and parallel to the yz plane (i.e. along RS). Each one will have a gradient.

This need for two rates of change leads us to the definition of the *partial derivatives* of z with respect to x and with respect to y, respectively denoted by

$$\frac{\partial z}{\partial x} \quad \text{and} \quad \frac{\partial z}{\partial y}$$

the ∂/∂ notation is to distinguish partial derivatives from 'full' derivatives, such as dy/dx.

Thus, $\partial z/\partial x$ is the partial derivative or rate of change of z with respect to x *while y remains constant* (i.e. the gradient of section PQ). We find it using the normal rule but treating y as if it were no different from any other constant (e.g. 3 or $17\frac{1}{2}$). For example,

$$\text{if } z = 5x^2y^3$$

we differentiate z w.r.t. (with respect to) x as if $5y^3$ were a constant multiplying x^2;

$$\text{then } \frac{\partial z}{\partial x} = 5 \cdot 2x \cdot y^3$$
$$= 10xy^3$$

Similarly, $\partial z/\partial y$ is the partial derivative or rate of change of z w.r.t. y while x remains constant (i.e. the gradient of Section RS). To find it we differentiate z w.r.t. y as if x were any constant; hence

$$\text{if } z = 5x^2y^3$$
$$\text{then } \frac{\partial z}{\partial y} = 5x^2 \cdot 3y^2$$
$$= 15x^2y^2$$

Example 1

Find $\partial z/\partial x$ and $\partial z/\partial y$ for the function

$$z = 2x^2y + xy - 1$$

Differentiating, we get

$$\frac{\partial z}{\partial x} = 2 \cdot 2x \cdot y + 1 \cdot y - 0$$
$$= 4xy + y$$
$$\frac{\partial z}{\partial y} = 2x^2 \cdot 1 + x \cdot 1 - 0$$
$$= 2x^2 + x$$

(Note: Constants always differentiate to zero w.r.t any variable.)

The two partial derivatives will in general be different.

The formal statement of them, just as we previously defined differentiation to be a limiting process, is

$$\frac{\partial z}{\partial x} = \lim_{h \to 0} \left\{ \frac{f(x+h, y) - f(x, y)}{h} \right\}$$

$$\frac{\partial z}{\partial y} = \lim_{h \to 0} \left\{ \frac{f(x, y+h) - f(x, y)}{h} \right\}$$

Notice how in each definition one variable stays fixed while the other changes.

Let us use the first of these to differentiate

$$z = 2x^2y + xy - 1$$

from first principles, in order to check that using the usual differentiation rule w.r.t. x together with the extra idea of y being constant, which we performed in Example 1 above, does give the same answer. To find $f(x + h, y)$ we just put $(x + h)$ instead of x everywhere that it occurs in $f(x,y)$:

$$\frac{\partial z}{\partial x} = \lim_{h \to 0} \left\{ \frac{[2(x+h)^2 y + (x+h)y - 1] - (2x^2y + xy - 1)}{h} \right\}$$

$$= \lim_{h \to 0} \left\{ \frac{[2(x^2 + 2xh + h^2)y + xy + hy - 1] - (2x^2y + xy - 1)}{h} \right\}$$

$$= \lim_{h \to 0} \left\{ \frac{2x^2y + 4xhy + 2h^2y + xy + hy - 1 - 2x^2y - xy + 1}{h} \right\}$$

$$= \lim_{h \to 0} \left\{ \frac{4xhy + 2h^2y + hy}{h} \right\}$$

$$= \lim_{h \to 0} \{4xy + 2hy + y\}$$

When $h = 0$

$$\frac{\partial z}{\partial x} = 4xy + y$$

which is what we obtained in Example 1. We find $\partial z/\partial y$ similarly, and you are asked to find it in Exercise 2.

Example 2

Find the two first-order derivatives (i.e. differentiate z once) $\partial z/\partial x$ and $\partial z/\partial y$ of

$$z = x^4 - 3x^3y^2 + 7x^2y - y + 15$$

Differentiating we get

$$\frac{\partial z}{\partial x} = 4x^3 - 3 \cdot 3x^2 \cdot y^2 + 7 \cdot 2x \cdot y - 0 + 0$$

$$= 4x^3 - 9x^2y^2 + 14xy$$

since y is treated like any other constant; and

$$\frac{\partial z}{\partial y} = 0 - 3x^3 \cdot 2y + 7x^2 \cdot 1 - 1 + 0$$

$$= -6x^3y + 7x^2 - 1$$

since x is treated like any other constant.

It is important to understand what the derivatives mean and to become familiar with their rules of operation, which are, after all, fairly simple. Nonetheless, you need them most as tools in subjects like Economics, where the variables are not usually called x and y but rather p_A, q_B, L, K, and so on. Try to adopt the approach that a tool like partial differentiation is most powerful when you can see its *general* application to a problem that only seems to be particular because of its notation and form of expression. If an economist presents you with a mathematical model, look at it with a mathematical eye to its form in terms of equations, rates of change or whatever. Apply your general mathematical principles, and literally take the problem apart.

Example 3
A production function is given by

$$Q = aL^b K^c$$

where Q is output, L is labour (man hours/week) and K is capital (£/week); a, b and c are constants, with

$$b + c = 1$$

Find $\partial Q/\partial L$ and $\partial Q/\partial K$.

We treat a, b and c like any constant numbers when applying the usual differentiation rule; and, of course, we also treat the variable not being differentiated as if it were a constant. Differentiating, we get

$$\frac{\partial Q}{\partial L} = abL^{b-1}K^c$$

We can improve this by multiplying by L/L (= 1); thus,

Partial Differential Calculus 57

$$\frac{\partial Q}{\partial L} = abL^{b-1}K^c \cdot \frac{L}{L}$$

$$= \frac{abL^b K^c}{L}$$

Substituting the original function gives

$$\frac{\partial Q}{\partial L} = \frac{bQ}{L}$$

Similarly, we get

$$\frac{\partial Q}{\partial K} = aL^b cK^{c-1}$$

$$= aL^b cK^{c-1} \cdot \frac{K}{K}$$

$$= \frac{aL^b cK^c}{K}$$

$$= \frac{cQ}{K}$$

If we were given a function with particular values of a, b and c, we could either substitute into the above or differentiate the function directly. The two partial derivatives are known as *marginal products*, analogously to our previous ideas of marginal revenue and marginal cost: $\partial Q/\partial L$ measures the rate of change of output w.r.t. labour (constant capital), and $\partial Q/\partial K$ measures the rate of change of output w.r.t. capital (labour constant).

Given particular values of L and K, we could easily substitute to find the values of $\partial Q/\partial L$ and $\partial Q/\partial K$.

Example 4

A production function is given by

$$F(L,K) = 2L^{1/3}K^{2/3}$$

Find $\partial F/\partial L$ and $\partial F/\partial K$, and evaluate them when $L = 8$ and $K = 27$. Also, if the output is fixed at 40, find the rate of change of L with K when $K = 20$.

Differentiating we get

$$F = 2L^{1/3}K^{2/3}$$

$$\frac{\partial F}{\partial L} = 2 \cdot \frac{L^{1/3-1}}{3} \cdot K^{2/3}$$

$$= 2 \cdot \frac{L^{-2/3}}{3} \cdot K^{2/3}$$

$$= \tfrac{2}{3} \cdot \frac{K^{2/3}}{L^{2/3}}$$

$$\frac{\partial F}{\partial K} = 2L^{1/3} \cdot \frac{2K^{2/3-1}}{3}$$

$$= 2L^{1/3} \cdot \frac{2K^{-1/3}}{3}$$

$$= \tfrac{4}{3} \cdot \frac{L^{1/3}}{K^{1/3}}$$

When $L = 8$ and $K = 27$,

$$\frac{\partial F}{\partial L} = \tfrac{2}{3} \cdot \frac{(27)^{2/3}}{(8)^{2/3}} = \tfrac{2}{3} \cdot \tfrac{9}{4} = \tfrac{18}{12} = \tfrac{3}{2}$$

$$\frac{\partial F}{\partial K} = \tfrac{4}{3} \cdot \frac{(8)^{1/3}}{(27)^{1/3}} = \tfrac{4}{3} \cdot \tfrac{2}{3} = \tfrac{8}{9}$$

The units of these answers would be something suitable like cases/man-hour or tons/£ employed, both over a certain time.

If the output is fixed at 40, the function becomes

$$40 = 2L^{1/3}K^{2/3}$$

Cancelling 2 and raising everything to power 3 give

$$20^3 = LK^2$$

$$L = \frac{20^3}{K^2}$$

$$= 20^3 K^{-2}$$

This is just a function of one variable, so we can find the rate of change of L w.r.t. K (i.e. dL/dK):

$$\frac{dL}{dK} = 20^3(-2K^{-3})$$

$$= -\frac{2 \cdot 20^3}{K^3}$$

When $K = 20$,

$$\frac{dL}{dK} = -\frac{2 \cdot 20^3}{20^3} = -2$$

That is, labour is being substituted by capital at the rate of -2 (man-hours/£ or whatever), which means a fall of 2 man-hours in labour used for every extra £1 of capital employed. Equivalently, every extra 2 man-hours in labour cause a fall of £1 in the capital employed.

Example 4 is just one of many similar ones that are possible, so remember to apply your general principles to the particular model with which you are faced.

The next analogy that we can draw from functions of one variable to functions of two variables is higher order differentiation (i.e. more than one). After differentiating

$$y = f(x)$$

to find dy/dx, we were able to differentiate again to find d^2y/dx^2, and this was important in finding the turning points (maximum and minimum points). Can we do something similar on the partial derivatives $\partial z/\partial x$ and $\partial z/\partial y$?

Since there are two partial derivatives, we can in fact differentiate each of them again w.r.t. both x and y, so we get *four* second-order derivatives. The two obvious ones are $\partial^2 z/\partial x^2$ and $\partial^2 z/\partial y^2$, where the function z has been differentiated twice w.r.t. one variable. The same rules as before apply. So, using the same function

we get
$$z = 2x^2y + xy - 1$$

$$\frac{\partial z}{\partial x} = 4xy + y \qquad\qquad \frac{\partial z}{\partial y} = 2x^2 + x$$

$$\frac{\partial^2 z}{\partial x^2} = 4y \qquad\qquad \frac{\partial^2 z}{\partial y^2} = 0$$

treating y as a constant   treating x as constant

The only thing to really watch for in the *mixed* second-order differentiation is the order in which the differentiations are done:

differentiating $\dfrac{\partial z}{\partial x}$ w.r.t. y gives $\dfrac{\partial^2 z}{\partial x \partial y}$ or $\dfrac{\partial}{\partial y}\left(\dfrac{\partial z}{\partial x}\right)$

differentiating $\dfrac{\partial z}{\partial y}$ w.r.t. x gives $\dfrac{\partial^2 z}{\partial y \partial x}$ or $\dfrac{\partial}{\partial x}\left(\dfrac{\partial z}{\partial y}\right)$

From the function

$$z = 2x^2 y + xy - 1$$

we get

$$\dfrac{\partial z}{\partial x} = 4xy + y \qquad \dfrac{\partial z}{\partial y} = 2x^2 + x$$

$$\dfrac{\partial^2 z}{\partial x \partial y} = 4x + 1 \qquad \dfrac{\partial^2 z}{\partial y \partial x} = 4x + 1$$

treating x as constant   treating y as constant

Notice that they are equal despite the different order of the differentiations w.r.t. x and y. Generally, they will be equal for all the functions that you meet.

Example 5
Find the four second-order derivatives of the function

$$f(x, y) = xy^2 + x^2 y^3 + 4x - y + 2$$

Differentiating, we get

$$\dfrac{\partial f}{\partial x} = 1 \cdot y^2 + 2x \cdot y^3 + 4$$

$$\dfrac{\partial^2 f}{\partial x^2} = 2 \cdot y^3 \qquad \dfrac{\partial^2 f}{\partial x \partial y} = 2y + 2x \cdot 3y^2$$

$$= 2y^3 \qquad\qquad\qquad = 2y + 6xy^2$$

and

$$\frac{\partial f}{\partial y} = x \cdot 2y + x^2 \cdot 3y^2 - 1$$

$$\frac{\partial^2 f}{\partial y \partial x} = 1 \cdot 2y + 2x \cdot 3y^2 \qquad \frac{\partial^2 f}{\partial y^2} = x \cdot 2 + x^2 \cdot 6y$$

$$= 2y + 6xy^2 \qquad\qquad = 2x + 6x^2 y$$

Again, note that $\partial^2 f/\partial x \partial y = \partial^2 f/\partial y \partial x$.

MAXIMUM AND MINIMUM OF FUNCTIONS OF TWO VARIABLES

In three dimensions a surface can have maxima and minima just as a curve can in two dimensions. We need the first derivative dy/dx to be zero for a horizontal gradient on the curve; then, in order to distinguish between maximum and minimum, we have a condition on the second derivative $d^2 y/dx^2$ that effectively tells us which way the curve is turning. A completely analogous result holds for a surface (i.e. a function of two variables). To demonstrate this clearly, let us first see what a maximum, a minimum and a new alternative called a *saddle-point* look like. We can then state the general analogous conditions needed for each and, finally, explain what these mean by reference to the diagrams in Figure 5.4.

The conditions for a function

$$z = f(x, y)$$

are:

(1) For all three cases

$$\frac{\partial z}{\partial x} = 0 = \frac{\partial z}{\partial y}$$

(2) Then, there is a saddle-point (see Figure 5.4c) if

$$\frac{\partial^2 z}{\partial x^2} \cdot \frac{\partial^2 z}{\partial y^2} - \left(\frac{\partial^2 z}{\partial x \partial y}\right)^2 < 0$$

(3) Or there is *either* a maximum *or* a minimum if

$$\frac{\partial^2 z}{\partial x^2} \cdot \frac{\partial^2 z}{\partial y^2} - \left(\frac{\partial^2 z}{\partial x \partial y}\right)^2 > 0$$

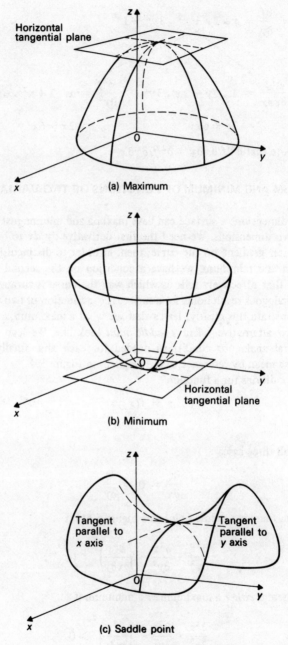

Figure 5.4

(4) Following (3), there is a maximum (see Figure 5.4a) if

$$\frac{\partial^2 z}{\partial x^2} < 0 \quad \text{and} \quad \frac{\partial^2 z}{\partial y^2} < 0$$

(5) Or there is a minimum (see Figure 5.4b) if

$$\frac{\partial^2 z}{\partial x^2} > 0 \quad \text{and} \quad \frac{\partial^2 z}{\partial y^2} > 0$$

Figure 5.4 shows that condition (1) is needed in all three cases. A tangential plane is horizontal to the 'peak' in (a) or to the 'trough' in (b). In (c) the two tangents are flat at the meeting of the directions going 'up then over' *and* 'down then up'. At all these three points it is therefore necessary that the rate of change of z w.r.t. x and w.r.t. y be zero. I shall not attempt to explain exactly what conditions (2) and (3) mean, but the offshoots (4) and (5) of (3) are similar to our previous one-variable condition to distinguish between max. and min. In (a) the rates of change of both first-order rates of change w.r.t. their own variable move from positive (slopes) to negative (i.e. decrease). In (b) the reverse is true.

Example 6
Find the points of relative maxima or minima (sometimes called the **critical points**) of the function

$$z = x^3 + 2y^3 - x^2 - 4y^2 - x - 8y$$

and distinguish between them.

Differentiating once, we get

$$\frac{\partial z}{\partial x} = 3x^2 - 2x - 1 \qquad \frac{\partial z}{\partial y} = 6y^2 - 8y - 8$$

At the critical points

$$\frac{\partial z}{\partial x} = 0 \qquad\qquad \frac{\partial z}{\partial y} = 0$$

$$3x^2 - 2x - 1 = 0 \qquad\qquad 6y^2 - 8y - 8 = 0$$
$$(3x + 1)(x - 1) = 0 \qquad\qquad 3y^2 - 4y - 4 = 0$$
$$(3y + 2)(y - 2) = 0$$
$$x = -\tfrac{1}{3} \text{ or } 1 \qquad\qquad y = -\tfrac{2}{3} \text{ or } 2$$

Differentiating for the second time, we get

$$\frac{\partial^2 z}{\partial x^2} = 6x - 2 \qquad \frac{\partial^2 z}{\partial y^2} = 12y - 8$$

$$\frac{\partial^2 z}{\partial x \partial y} = \frac{\partial^2 z}{\partial y \partial x} = 0$$

When $x = -\frac{1}{3}$ and $y = -\frac{2}{3}$,

$$\frac{\partial^2 z}{\partial x^2} = -4 < 0 \qquad \frac{\partial^2 z}{\partial y^2} = -16 < 0 \qquad \text{(condition 4)}$$

$$\frac{\partial^2 z}{\partial x^2} \cdot \frac{\partial^2 z}{\partial y^2} - \left(\frac{\partial^2 z}{\partial x \partial y}\right)^2 = (-4)(-16) - (0)^2 = 64 > 0 \text{ (condition 3)}$$

$$\therefore \text{ a maximum at } (-\tfrac{1}{3}, -\tfrac{2}{3}, z_{\max})$$

Substitution into z gives the actual value $z_{\max}$ of the relative maximum (i.e. the point at the peak in Figure 5.4a).

When $x = 1$ and $y = 2$,

$$\frac{\partial^2 z}{\partial x^2} = 4 > 0 \qquad \frac{\partial^2 z}{\partial y^2} = 16 > 0 \quad \text{(condition 5)}$$

$$\frac{\partial^2 z}{\partial x^2} \cdot \frac{\partial^2 z}{\partial y^2} - \left(\frac{\partial^2 z}{\partial x \partial y}\right)^2 = (4)(16) - (0)^2 = 64 > 0 \quad \text{(condition 3)}$$

$$\therefore \text{ a minimum at } (1, 2, z_{\min})$$

Similarly, substitution into z gives the point at the trough in Figure 5.4b.

Note that condition 3 is not satisfied either when $x = -\frac{1}{3}$ and $y = 2$ or when $x = 1$ and $y = -\frac{2}{3}$. Therefore, neither a maximum nor a minimum occurs at these points.

Example 7

A manufacturer produces two versions of his product: the basic B and the super S. The selling prices of each are given by

$$p_B = 12 - 3q_B$$
$$p_S = 10 - q_S$$

where q_B and q_S are the quantities of each. The joint total-cost function is given by

$$TC = q_B^2 + 2q_B q_S + q_S^2$$

Find the levels of production required to maximise the joint profit and the value of the maximum joint profit.

Notice the realistic demand functions and the way in which the quantities of each version of the product are inter-related in their effect on costs. The examples that you are likely to meet in Economics may express the connections of prices, quantities, revenues, costs, and so on in various ways. Always read the question carefully and construct *first* the functions of total revenue, total cost or whatever is needed. *Then* apply the mathematics, differentiating w.r.t. the two variables in which your function is *finally* expressed (usually two quantities *or* two prices).

Total revenue is given by

$$TR = p_B q_B + p_S q_S$$
$$= (12 - 3q_B)q_B + (10 - q_S)q_S$$
$$= 12q_B - 3q_B^2 + 10q_S - q_S^2$$

With total cost as given above:

$$TC = q_B^2 + 2q_B q_S + q_S^2$$

total profit is given by

$$TP = TR - TC$$
$$= (12q_B - 3q_B^2 + 10q_S - q_S^2) - (q_B^2 + 2q_B q_S + q_S^2)$$
$$= 12q_B + 10q_S - 2q_B q_S - 4q_B^2 - 2q_S^2$$

Differentiating, we get

$$\frac{\partial TP}{\partial q_B} = 12 - 2q_S - 8q_B \qquad \frac{\partial TP}{\partial q_S} = 10 - 2q_B - 4q_S$$

Putting these first-order derivatives equal to zero gives

$$12 - 2q_S - 8q_B = 0 \qquad 10 - 2q_B - 4q_S = 0$$
$$6 - q_S - 4q_B = 0 \qquad 5 - q_B - 2q_S = 0$$

To solve these equations we can simply substitute (or solve simultaneously); hence,

$$6 - q_S - 4(5 - 2q_S) = 0$$
$$6 - q_S - 20 + 8q_S = 0$$
$$7q_S = 14$$
$$q_S = 2$$

substituting back gives

$$5 - q_B - 2(2) = 0$$
$$q_B = 1$$

Differentiating for the second time gives

$$\frac{\partial^2 TP}{\partial q_B^2} = -8 \qquad \frac{\partial^2 TP}{\partial q_S^2} = -4$$

$$\frac{\partial^2 TP}{\partial q_B \partial q_S} = -2 = \frac{\partial^2 TP}{\partial q_S \partial q_B}$$

Hence

$$\frac{\partial^2 TP}{\partial q_B^2} \cdot \frac{\partial^2 TP}{\partial q_S^2} - \left(\frac{\partial^2 TP}{\partial q_B \partial q_S}\right)^2 = (-8)(-4) - (-2)^2 = 28 > 0$$

∴ a *maximum* where $q_B = 1, q_S = 2$

When $q_B = 1$ and $q_S = 2$,

$$p_B = 9 \qquad p_S = 8$$

Hence,

$$TP_{max} = [(9)(1) + (8)(2)] - [1^2 + (2)(1)(2) + 2^2] = 25 - 9 = 16$$

That is, a maximum joint profit of 16 is made when one basic and two super are produced. (All prices and quantities will, of course, be in some suitable units.)

USE OF LAGRANGE MULTIPLIER

We shall conclude by considering a slightly more complicated case of the above, namely, when we wish to maximise or minimise a function *subject to a constraint*. In real life, options may be limited by some restriction on the quantities produced (because of an order) or on the total cost (because of a budget). We can solve this type of problem in a very similar way to before, but the background mathematically is very difficult without further knowledge. Nonetheless, the method is easy, so just follow through the working of the next example.

Example 8
Mowclose Ltd markets two types of lawnmower: the Ace and the Champion, which cost £10 and £15 per unit respectively to produce. There is a budget constraint of £850. The profit function is considered to be

$$60x + 150y - x^2 - 3y^2$$

where x and y are the numbers of Ace and Champion lawnmowers respectively. What values of x and y will maximise the profit?

The budget constraint on total cost is

$$10x + 15y = 850$$

The profit function given could be maximised alone if it were not for the constraint. We have to alter it by adding to it an unknown multiple λ of the left hand side of the constraint function in its form

$$g(x, y) = 0$$

That is, we add $\lambda(10x + 15y - 850)$ to the profit function. We make the assumption that this new compound function has its maximum at the same point as the profit function. Thus, we have a new compound function F where

$$F = 60x + 150y - x^2 - 3y^2 + \lambda(10x + 15y - 850)$$
$$= 60x + 150y - x^2 - 3y^2 + 10\lambda x + 15\lambda y - 850\lambda$$

We now carry out exactly the same procedure as before, but differentiating w.r.t. x, y and λ, because the *extra* thing that we need is the value of this multiple λ (called the *Lagrange multiplier*) which gives F its maximum. Thus,

$$\frac{\partial F}{\partial x} = 60 - 2x + 10\lambda$$

$$\frac{\partial F}{\partial y} = 150 - 6y + 15\lambda$$

$$\frac{\partial F}{\partial \lambda} = 10x + 15y - 850$$

Equating these all to zero, as before, gives

$$60 - 2x + 10\lambda = 0$$
$$150 - 6y + 15\lambda = 0$$
$$10x + 15y - 850 = 0 \text{ (the constraint function again)}$$

The first two yield

$$30 + 5\lambda = x$$
$$\tfrac{1}{2}(50 + 5\lambda) = y$$

So, substituting into the third, we get

$$10(30 + 5\lambda) + \tfrac{15}{2}(50 + 5\lambda) - 850 = 0$$
$$20(30 + 5\lambda) + 15(50 + 5\lambda) - 1{,}700 = 0$$
$$4(30 + 5\lambda) + 3(50 + 5\lambda) - 340 = 0$$
$$120 + 20\lambda + 150 + 15\lambda - 340 = 0$$
$$35\lambda - 70 = 0$$
$$35\lambda = 70$$
$$\lambda = 2$$

Substituting back gives

$$x = 30 + 5(2) = 40$$
$$y = \tfrac{1}{2}[50 + 5(2)] = 30$$

That is, Mowclose needs to make 40 Ace mowers and 30 Champion mowers to maximise its profit. Substitution of these numbers into the profit function will give the value of the maximum profit.

An alternative method, which incidentally checks our answer and shows that the Lagrange multiplier method is valid, is to substitute from the constraint function into the profit function and then to maximise the resulting function of one variable. Thus, from the constraint function we get

$$x = \frac{850 - 15y}{10}$$
$$= \frac{170 - 3y}{2}$$

which we substitute into the profit function as follows:

$$P = 60x + 150y - x^2 - 3y^2$$

$$= 60\left(\frac{170 - 3y}{2}\right) + 150y - \left(\frac{170 - 3y}{2}\right)^2 - 3y^2$$

$$= 30(170 - 3y) + 150y - \tfrac{1}{4}(170^2 - 1{,}020y + 9y^2) - 3y^2$$

$$= 5{,}100 - 90y + 150y - \tfrac{1}{4}(170)^2 + 255y - \tfrac{9}{4}y^2 - 3y^2$$

We can avoid simplifying, because differentiation removes some terms:

$$\frac{dP}{dy} = -90 + 150 + 255 - \tfrac{9}{2}y - 6y$$

$$= 315 - \tfrac{21}{2}y$$

At turning points

$$315 - \tfrac{21}{2}y = 0$$

$$y = 30 \quad \text{(as before)}$$

Substituting back gives

$$x = 40 \quad \text{(as before)}$$

Also

$$\frac{d^2P}{dy^2} = -\tfrac{21}{2} < 0$$

∴ a *maximum* when $x = 40$, $y = 30$ (as before)

FURTHER EXAMPLES OF USE OF LAGRANGE MULTIPLIER

Suppose that we have a general production function (output Q)

$$Q = aL^b K^c$$

and a general cost function (cost C)

$$C = rL + sK$$

where the variables L and K represent labour and capital respectively and a, b, c, r and s are simply constants.

We shall consider two examples:

(1) minimising cost C subject to a constraint u on output (Example 9); and
(2) maximising output Q subject to a constraint v on cost (Example 10).

Example 9
Minimise $(rL + sK)$ subject to

$$aL^b K^c - u = 0$$

Using the Lagrange multiplier λ, we seek to minimise

$$F = rL + sK + \lambda(aL^b K^c - u)$$

Differentiating F partially w.r.t. L, K and λ, we get

$$\frac{\partial F}{\partial L} = r + \lambda ab L^{b-1} K^c$$

$$\frac{\partial F}{\partial K} = s + \lambda ac L^b K^{c-1}$$

$$\frac{\partial F}{\partial \lambda} = aL^b K^c - u$$

All these first-order partial derivatives must equal zero for max. or min.; hence,

$$r = -\lambda ab L^{b-1} K^c \tag{1}$$

$$s = -\lambda ac L^b K^{c-1} \tag{2}$$

$$aL^b K^c = u \tag{3}$$

Dividing (1) by (2) gives

$$\frac{r}{s} = \frac{b}{c} \cdot \frac{K}{L}$$

$$K = \frac{rcL}{sb}$$

Substituting into (3) gives

$$aL^b \left(\frac{rcL}{sb}\right)^c = u$$

$$\therefore L^{b+c} = \text{some constant} \quad (= L \text{ if } b+c = 1)$$

Thus, L can be found, and therefore K, when the constants are given numerical values. Hence, we find $C_{\min}$.

Example 10
Maximise $aL^b K^c$ subject to

$$rL + sK - v = 0$$

Using the Lagrange multiplier λ, we seek to maximise

$$G = aL^b K^c + \lambda(rL + sK - v)$$

Differentiating G partially w.r.t. L, K and λ, we get

$$\frac{\partial G}{\partial L} = abL^{b-1}K^c + \lambda r$$

$$\frac{\partial G}{\partial K} = acL^b K^{c-1} + \lambda s$$

$$\frac{\partial G}{\partial \lambda} = rL + sK - v$$

All these first-order partial derivatives must equal zero for max. or min., hence,

$$\lambda r = -abL^{b-1}K^c \tag{4}$$

$$\lambda s = -acL^b K^{c-1} \tag{5}$$

$$rL + sK = v \tag{6}$$

Dividing (4) by (5) gives

$$\frac{r}{s} = \frac{b}{c} \cdot \frac{K}{L}$$

$$K = \frac{rcL}{sb} \quad \text{(as before)}$$

Substituting into (6) gives

$$rL + \not{s} \cdot \frac{rcL}{\not{s}b} = v$$

$$L\frac{(b+c)}{b}r = v$$

$$\therefore \; L = \frac{bv}{r} \quad \text{if } b+c = 1$$

Substituting back, we get

$$K = \frac{\not{r}c}{s\not{b}} \cdot \frac{\not{b}v}{\not{r}}$$

$$= \frac{cv}{s}$$

Hence, we find $Q_{\max}$.

One fear that I have in using an example to introduce a method is that you may lose sight of the general situation among the particular figures of the example. I shall therefore state the formal procedure of the Lagrange multiplier method in the general case as follows:

Given an objective function

$$z = f(x, y)$$

subject to the constraint

$$g(x, y) = 0$$

the (compound) Lagrange function F is constructed as

$$F = f(x, y) + \lambda g(x, y)$$

Then, for a stationary (max., min. or saddle-point) value of F we need the conditions

$$\frac{\partial F}{\partial x} = \frac{\partial F}{\partial y} = \frac{\partial F}{\partial \lambda} = 0$$

EXERCISES

(1) Evaluate the following functions at the point $(-1, 2)$:

 (a) $f(x, y) = x + 2$.
 (b) $f(x, y) = 2xy + 3$.
 (c) $f(x, y) = 3x^2 y^2 - x + 1$.
 (d) $f(x, y) = xe^{y+2}$.

(2) Differentiate

$$z = 2x^2 y + xy - 1$$

partially w.r.t. y from the first-principles definition, and check the answer obtained in Example 1 with the normal rule.

(3) Find $\partial z / \partial x$ and $\partial z / \partial y$ for the function

$$z = 4x^4 y^2 - 3x^2 + xy^3 + 5 + y - 11$$

(4) Given the production function

$$P(L, K) = 3L^{1/2} K^{1/2}$$

find $\partial P / \partial L$ and $\partial P / \partial K$ when $L = 400$ and $K = 25$.

(5) A firm makes two types of product, which have joint cost function

$$C = 2x + \tfrac{1}{100}y^2 + y + 125$$

where x and y are the numbers of each type of product made. Find $\partial C/\partial x$ and $\partial C/\partial y$ when $x = 10$ and $y = 20$.

(6) The demand functions for the two types A and B of a manufacturer's product are

$$q_A = 500 - 20p_A + p_B$$
$$q_B = 250 + 2p_A - 9p_B$$

Find

$$\frac{\partial q_A}{\partial p_A}, \quad \frac{\partial q_A}{\partial p_B}, \quad \frac{\partial q_B}{\partial p_A} \quad \text{and} \quad \frac{\partial q_B}{\partial p_B}$$

(7) Find the four second-order partial derivatives of the function

$$z = (x + y)(x^2 + y^2 - x)$$

(8) Investigate fully the critical points of the function

$$z = x^2 + y^3 - 4x - 3y + 5$$

Find the values of any relative maxima or minima.

(9) Two competing soapbars are manufactured by a monopolist. Bar X costs 9p each to produce and sells at p_x pence, with the demand (in hundreds/day) given by

$$x = 2(p_y - p_x) + 4$$

Bar Y costs 12 pence each to produce and sells at p_y pence, with the demand (in hundreds/day) given by

$$y = \frac{p_x}{4} - \frac{5p_y}{2} + 52$$

Show that maximum joint profit is achieved with prices of $p_x = 16$ pence and $p_y = 20$ pence.

(10) The demand functions for two products A and B are given by

$$p_A = 45 - q_A^2 - q_A$$
$$p_B = 20 - q_B^2 - q_B$$

The cost functions are given by $2q_A^2$ and q_B^2 respectively. Find the values of q_A and q_B that maximise the joint profit.

(11) A. Warr Toys Ltd markets the Aggroman doll and the Actionpack accessory. The cost of making them is 60p per doll and 8p per accessory. On the assumption that there is no competition in this market sector, the demand function for the doll is estimated at

$$x_d = 330 - 5p_d - 2p_a$$

and the demand function for the accessory is estimated at

$$x_a = 200 - 2p_d - \tfrac{5}{2}p_a$$

Find the selling prices of doll and accessory that maximise the joint profit.

Comment on the answers.

(12) A firm operates with a total cost function of

$$3q_1^2 + q_1q_2 + 2q_2^2 + 500$$

where q_1 and q_2 are the quantities of units produced in the firm's two factories in Enfield and Basildon respectively. If the firm has to fulfil an order for 400 units, how should it distribute production between the two factories so as to minimise its costs?

6 Integral Calculus

INTEGRATION

We have already tackled the subject of Differentiation in Chapters 4 and 5. For a function of one variable, such as

$$y = -3 + 4x - x^2$$

we saw that the differential or derivative of y w.r.t. x, denoted by dy/dx ($= 4 - 2x$ by the usual rule), represents the rate at which y is changing w.r.t. x. This differential is also a function of x, and substitution of a particular value for x will give the rate of change of y w.r.t. x at that value.

In Figure 6.1,

$$y = -3 + 4x - x^2$$

$$\frac{dy}{dx} = 4 - 2x$$

At A, where $x = 1$,

$$\frac{dy}{dx} = 4 - 2(1) = 2$$

$\therefore$ tangent at A has gradient 2

At B, where $x = 2$,

$$\frac{dy}{dx} = 4 - 2(2) = 0$$

$\therefore$ tangent at B is horizontal

This process of differentiating the function to get its rate of change is properly called an *operator*. Can we *reverse* the operator to obtain the function from its rate of change? There is no logical reason to suppose that we cannot; and since a revenue function, for example, can be differentiated to give the marginal revenue function, it should be possible to reverse the process to obtain the revenue function from a marginal revenue

76 The Readable Maths and Statistics Book

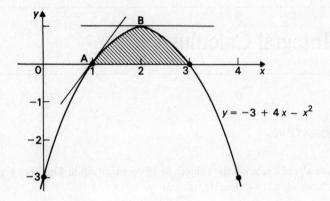

Figure 6.1

function. This reverse process of differentiating we call *integration*. Thus, the operators and their symbols can be shown as

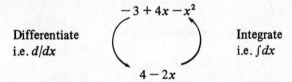

Differentiate
i.e. d/dx

Integrate
i.e. $\int dx$

Our next logical thought is that this operator (i.e. process of integrating) should have a rule similar to the rule for differentiating. In fact the general rule is

$$\text{if} \quad y = ax^n$$

$$\text{then} \quad \int y\,dx = \int ax^n dx = \frac{ax^{n+1}}{n+1} + c$$

where c is a constant.

Remember that $\int dx$ merely tells you what to do. It is the symbol of the integrating operator, just as d/dx is the symbol of the differentiating operator. The rule is true for all values of n (positive, negative or fractional) *except $n = -1$* (see Example 9).

Notice that the rule contains an arbitrary constant c. Let us check that this rule does in fact reverse the operation for

$$y = -3 + 4x - x^2$$

Integral Calculus

and, in doing so see why the constant is necessary. Having differentiated once to get $(4-2x)$, we apply the integration rule to get

$$\int(4-2x)dx = \frac{4x}{1} - \frac{2x^{1+1}}{1+1} + c$$

$$= 4x - \frac{2x^2}{2} + c$$

$$= 4x - x^2 + c$$

(Remember that 4 can be thought of as $4x^0$ and $-2x$ as $-2x^1$.) Have we come back to the original function $(-3 + 4x - x^2)$? Well, yes we have, if the constant c is -3. The crucial point is that all constants differentiate to zero; so whatever constant had been present in the original function, it would have disappeared with the operation of differentiation. When we come to reverse the operation, we simply don't know, without further information, whether a constant was there or not. So, we put one in, its value being unknown but possibly zero to cover the case of an original function without a constant.

Example 1
Integrate the function

$$y = 2x^3$$

Using the integrating rule, we get

$$I = \int y\,dx = \int 2x^3\,dx$$

$$= \frac{2x^{3+1}}{3+1} + c$$

$$= \frac{2x^4}{4} + c$$

$$= \frac{x^4}{2} + c$$

We can use the differentiating rule to check the answer, since

$$\frac{dI}{dx} = \frac{d}{dx}\left(\int y\,dx\right) \quad \text{should} = y$$

as the operators should cancel. In this example

$$\frac{dI}{dx} = \frac{d}{dx}\left(\frac{x^4}{2} + c\right)$$

$$= \frac{4x^3}{2} + 0$$

$$= 2x^3$$

$$= y$$

$$\therefore I \text{ is correct}$$

The constant c remains unknown without further information, so we have what is called an *indefinite integral I*.

Example 2
Integrate the function

$$y = \frac{7}{x^2} + 3x - 4$$

Using the integrating rule, we get

$$I = \int y\,dx = \int\left(\frac{7}{x^2} + 3x - 4\right)dx$$

$$= \int(7x^{-2} + 3x - 4)dx$$

$$= \frac{7x^{-2+1}}{-2+1} + \frac{3x^{1+1}}{1+1} - \frac{4x^1}{1} + c$$

$$= \frac{7x^{-1}}{-1} + \frac{3x^2}{2} - 4x + c$$

$$= -\frac{7}{x} + \frac{3x^2}{2} - 4x + c$$

Example 3
Integrate $(-3 + 4x - x^2)$.

By the rule, and remembering, of course, that + and − signs merely separate the terms being operated on, we get

$$I = \int (-3 + 4x - x^2)dx$$

$$= -3x + \frac{4x^2}{2} - \frac{x^3}{3} + c$$

$$= -3x + 2x^2 - \frac{x^3}{3} + c$$

Again, differentiation would check the answer.

What we have actually done in Example 3 is to integrate the function $(-3 + 4x - x^2)$ (see Figure 6.1), whose rate of change we found to be $(-2x + 4)$ by differentiation. Does this integration of $(-3 + 4x - x^2)$ to obtain $(-3x + 2x^2 - \frac{1}{3}x^3 + c)$ yield anything meaningful? In fact it yields an area; and by substituting two particular values of x, we can find the particular area lying under the curve between these values of x (see the shaded area under the curve between $x = 1$ and $x = 3$ in Figure 6.1). We do the substitution as follows:

$$I = \int_1^3 (-3 + 4x - x^2)dx$$

$$= \left[-3x + 2x^2 - \frac{x^3}{3} + c \right]_1^3$$

$$= \left(-3(3) + 2(3)^2 - \frac{(3)^3}{3} + c \right) - \left(-3(1) + 2(1)^2 - \frac{(1)^3}{3} + c \right)$$

$$= (-9 + 18 - 9 + c) - (-3 + 2 - \tfrac{1}{3} + c)$$

$$= 0 + \cancel{c} + \tfrac{4}{3} - \cancel{c}$$

$$= \tfrac{4}{3}$$

That is, the shaded area is $\tfrac{4}{3}$ square units.

Notice how the integrating *limits* of 3 and 1, which simply tell us where the area is required to start and end, are placed on the integral sign $\int$ and are then substituted separately, before we subtract the lower limit substitution from the upper to get the answer. The constant c has to cancel, so we have what is called a *definite integral*.

Incidentally, the curve is a parabola and the area of $\tfrac{4}{3}$ square units represents two-thirds of the enclosing rectangle's area of 2 square units (see Figure 6.2). The ancient Greeks discovered this result for the parabola,

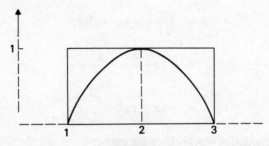

Figure 6.2

without knowing any integration, by weighing a rectangle of wood, drawing a parabola on it, then cutting out and weighing the parabola.

You have had to take my word for it until now that integration does yield an area, so let us work another example of a simpler case to show this. A rigorous explanation is given at the end of this chapter.

Example 4

Find the area under the function

$$y = 2 + \tfrac{3}{2}x$$

between $x = 2$ and $x = 6$ (see Figure 6.3).

Integration gives the area A under the line:

$$A = \int_2^6 y\,dx = \int_2^6 \left(2 + \frac{3x}{2}\right)dx$$

$$= \left[2x + \frac{3x^2}{2(2)}\right]_2^6 \quad (c \text{ not necessary})$$

$$= \left(2(6) + \frac{3(6)^2}{4}\right) - \left(2(2) + \frac{3(2)^2}{4}\right)$$

$$= (12 + 27) - (4 + 3)$$

$$= 39 - 7$$

$$= 32$$

We can also find the area using the usual trapezium formula or by considering A as a triangle on a rectangle. We get

$$A = \tfrac{1}{2}(5 + 11)(4) = 32 \quad \text{(as before)}$$

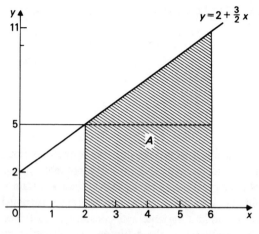

Figure 6.3

It is always useful to have a diagram, because we can only substitute limits relating to an area that is wholly one side of the x axis or the other. Failure to ensure this means that parts of the areas above and below the x axis interfere with one another, giving an incorrect answer. When the condition is met correctly, a positive answer comes from an area above the x axis and a negative answer from an area below the x axis (see Exercise 3).

There is another way in which the arbitrary constant c can be found apart from the area case above: when we have the equivalent of limits known as *boundary conditions* (see Example 5).

Example 5
Find the integral I of the function

$$y = 2x^4 - 3x^2 + 2$$

given that $I = 3$ when $x = 1$.

Integrating gives

$$I = \int y\,dx = \int (2x^4 - 3x^2 + 2)dx$$

$$= \frac{2x^5}{5} - \frac{3x^3}{3} + 2x + c$$

$$= \frac{2x^5}{5} - x^3 + 2x + c$$

Substituting $x = 1$ and $I = 3$ gives

$$3 = \frac{2(1)^5}{5} - (1)^3 + 2(1) + c = \tfrac{7}{5} + c$$

$$c = 3 - \tfrac{7}{5} = \tfrac{8}{5}$$

$$\therefore I = \frac{2x^5}{5} - x^3 + 2x + \tfrac{8}{5}$$

In our later work on distributions (e.g. the normal distribution) we shall see that an area under a curve can represent a probability. For example, we find the probability of having an IQ higher than a certain number, given the mean and standard deviation of the (normal) distribution of IQ, by converting to a standard variable and consulting the tables. These areas (which are probabilities) in the table are calculated by integration. In Chapter 16 we shall discuss probability density functions and the cumulative probability (i.e. area) underneath it, which we call the *distribution function $F(x)$*.

Following Figure 16.4 we have

$$F(x) = p(x \leqslant x)$$

which is a sum of (discrete) probabilities. Example 6 shows how this works for a continuous distribution.

Example 6
For the function

$$y = 6x - x^2$$

defined for $0 \leqslant x \leqslant 6$, find the probability that x has a value between 4 and 6.

We must, of course, draw the function first (see Figure 6.4). The total area A under the curve is given by

$$A = \int_0^6 y\,dx = \int_0^6 (6x - x^2)\,dx$$

$$= \left[\frac{6x^2}{2} - \frac{x^3}{3}\right]_0^6$$

$$= \left[3x^2 - \frac{x^3}{3}\right]_0^6$$

$$= \left(3(6)^2 - \frac{(6)^3}{3}\right) - (0)$$

$$= 108 - 72$$

$$= 36$$

The shaded area A_s between $x = 4$ and $x = 6$ is given by the same integral with different limits:

$$A_s = \left[3x^2 - \frac{x^3}{3}\right]_4^6$$

$$= \left(3(6)^2 - \frac{(6)^3}{3}\right) - \left(3(4)^2 - \frac{(4)^3}{3}\right)$$

$$= (108 - 72) - (48 - \tfrac{64}{3})$$

$$= 36 - \tfrac{80}{3}$$

$$= \tfrac{28}{3} \text{ or } 9\tfrac{1}{3}$$

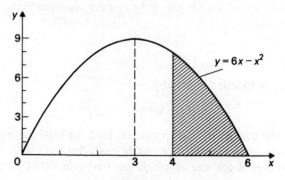

Figure 6.4

The probability required is given by

$$p = \frac{A_s}{A}$$

$$= \frac{9\frac{1}{3}}{36} = \frac{28}{108} = \frac{7}{27} \text{ or } 0.259$$

Notice that the y scale can easily be altered to make what is at present a frequency into a probability, simply by dividing by 36 to make the total area A have a value of 1. The function then becomes the probability density function $(6x - x^2)/36$, and the cumulative probabilities are given by the distribution function

$$F(b) = \int_0^b \left(\frac{6x - x^2}{36}\right) dx$$

where b is the value of x to which we want to go up. Check that a value of $b = 4$ gives

$$F(4) = \tfrac{20}{27}$$

ECONOMICS APPLICATIONS

There are no direct applications of integration to Economics in the same way as we have found for differentiation, but indirectly integration has an important role through its use in statistics and probability. What I shall do is to show several typical uses in Economics situations in the examples that follow.

Example 7
A marginal cost function is given by

$$MC = 3(5 + 6x - x^2)$$

where x is the number of units produced. Find the total cost and average cost functions, given that the total cost of producing one unit is 23.

Marginal cost is the *rate* at which the total cost TC of production is changing when the quantity produced changes; that is,

$$MC = \frac{dTC}{dx}$$

$$\therefore TC = \int MC\,dx = \int\left(\frac{dTC}{dx}\right)dx$$

$$= \int 3(5 + 6x - x^2)\,dx$$

$$= \int (15 + 18x - 3x^2)\,dx$$

$$= 15x + \frac{18x^2}{2} - \frac{3x^3}{3} + c$$

$$= 15x + 9x^2 - x^3 + c$$

Given (boundary conditions) $TC = 23$ when $x = 1$,

$$23 = 15(1) + 9(1)^2 - (1)^3 + c$$

$$= 23 + c$$

$$c = 0$$

$$\therefore TC = 15x + 9x^2 - x^3$$

Hence, average cost is given by

$$AC = \frac{TC}{x}$$

$$= \frac{1}{x}(15x + 9x^2 - x^3)$$

$$= 15 + 9x - x^2$$

Example 8

The marginal revenue function for an article of clothing is

$$MR = 3 - \frac{2}{\sqrt{x}}$$

where x is the quantity demanded. Find the total revenue function and the average revenue function (i.e. the 'demand function') if it is known that a demand for 100 articles gives a sales revenue of 200. Also, what sales revenue will be generated by a demand of 225 articles?

Marginal Revenue is the rate at which total revenue TR is changing when the quantity demanded changes; that is,

$$MR = \frac{dTR}{dx}$$

$$\therefore TR = \int MR\,dx = \int \left(\frac{dTR}{dx}\right)dx$$

$$= \int \left(3 - \frac{2}{\sqrt{x}}\right)dx$$

$$= \int (3 - 2x^{-1/2})dx \quad (\text{since } \sqrt{x} = x^{1/2})$$

$$= 3x - \frac{2x^{-1/2+1}}{-\tfrac{1}{2}+1} + c$$

$$= 3x - \frac{2x^{1/2}}{\tfrac{1}{2}} + c$$

$$= 3x - 4x^{1/2} + c$$

Given $TR = 200$ when $x = 100$,

$$200 = 3(100) - 4(100)^{1/2} + c$$

$$= 300 - 4(10) + c$$

$$= 260 + c$$

$$c = 200 - 260$$

$$= -60$$

$$\therefore TR = 3x - 4x^{1/2} - 60$$

Hence, average revenue is given by

$$AR = \frac{TR}{x}$$

$$= \frac{1}{x}(3x - 4x^{1/2} - 60)$$

$$= 3 - \frac{4}{x^{1/2}} - \frac{60}{x}$$

When $x = 225$,

$$TR = 3(225) - 4(225)^{1/2} - 60$$

$$= 675 - 4(15) - 60$$

$$= 555$$

Example 9

The Pseudo Cosmetics Company markets a new toiletry called Grit. Its economist estimates that Grit will capture a market (percentage) share after t years of

$$24\left(1 - \frac{1}{1+t}\right)$$

Draw the graph of this function from $t = 0$ to $t = 5$. Also, find the total income from Grit over the five-year period, at current prices, assuming a stable market sector with total expenditure E per annum.

The function is shown in Figure 6.5. The important factor here is that the market share s is considered to be growing continuously. The model is a realistic one because, after an initially steep rise, the rate of change of percentage share with time flattens off.

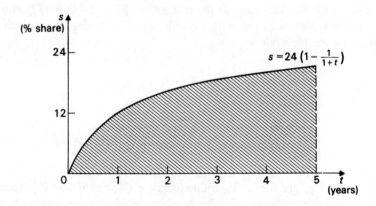

Figure 6.5

The fraction of the total expenditure captured by Grit at any time t will be

$$\frac{24}{100}\left(1 - \frac{1}{1+t}\right) \quad \text{(i.e. the due percentage)}$$

We want to know the *total* expenditure on Grit over the period $t = 0$ to $t = 5$, so we need the shaded area under the curve (i.e. the total market share over that period) times E:

$$\text{Income} = I = \int_0^5 \frac{24}{100}\left(1 - \frac{1}{1+t}\right)dt \cdot E$$

Notice that the letter given to the variable is unimportant in a definite integral, since the answer will be a number. Such a variable is called a *dummy variable*.) Any multiplying constant can be taken out as a factor; hence,

$$I = \frac{24E}{100} \int_0^5 \left(1 - \frac{1}{1+t}\right)dt$$

$$= \frac{24E}{100} \left[t - \log_e(1+t)\right]_0^5$$

since the integral of $1/(1+t)$ is $\log_e(1+t)$.

Note: This is the special case of x^n with $n = -1$, which we mentioned earlier as the one exception to the rule, since $1/(1+t) = (1+t)^{-1}$. Now $(1+t)$ is a linear function, with t coefficient of 1 obeying the same rules as x. The complete rule is

$$\int x^{-1} dx = \int \left(\frac{1}{x}\right) dx = \log_e x + c$$

$$\int (1+x)^{-1} dx = \int \left(\frac{1}{1+x}\right) dx = \log_e(1+x) + c$$

Note that $\log_e$ just means 'logarithm to base e' (instead of base 10). There is a $\log_e$ table in the set of ordinary tables, which you can look up very easily.

To continue the example:

$$I = \frac{24E}{100}\{[5 - \log_e (1+5)] - (-\log_e 1)\}$$

$$= \frac{6E}{25}(5 - \log_e 6) \quad \text{(as } \log_e 1 = 0)$$

$$= \frac{6E}{25}(5 - 1.7918)$$

$$= 0.77E$$

Example 10

The machines in a factory produce items that generate net revenue (£'000) of

$$N = 200 - \frac{t^2}{2}$$

where t is the time in years after the machines were new. The cost (also £'000) of maintaining the machines is given by

$$M = 50 + t^2$$

Sketch both functions on the same graph, and find:

(a) the value of t where net revenue is equal to maintenance cost.
(b) the total value of net profit over that period.

Once again the changes are taking place continuously, the net revenue gradually falling and the maintenance cost steadily rising. This is shown in Figure 6.6.

(a) When $N = M$,

$$200 - \frac{t^2}{2} = 50 + t^2$$

$$150 = \frac{3t^2}{2}$$

$$t^2 = 100$$

$$t = 10 \text{ years}$$

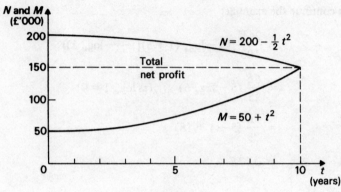

Figure 6.6

(b)

$$\text{Total } N = \int_0^{10} N\,dt = \int_0^{10}\left(200 - \frac{t^2}{2}\right)dt \quad \text{(i.e. area under } N\text{)}$$

$$= \left[200t - \frac{t^3}{6}\right]_0^{10}$$

$$= \left(2{,}000 - \frac{1{,}000}{6}\right) - (0)$$

$$= 1{,}833\tfrac{1}{3}$$

$$\text{Total } M = \int_0^{10} M\,dt = \int_0^{10}(50 + t^2)\,dt \quad \text{(i.e. area under } M\text{)}$$

$$= \left[50t + \frac{t^3}{3}\right]_0^{10}$$

$$= \left(500 + \frac{1{,}000}{3}\right) - (0)$$

$$= 833\tfrac{1}{3}$$

Total net profit = Total N − Total M (i.e. difference of areas)
= $1,833\frac{1}{3} - 833\frac{1}{3}$
= $1,000$ (i.e. £1,000,000)

Instead of subtracting *after* integration as above to find the total net profit, we could have found the net profit function as $(N-M)$ and then integrated it to get the same total net profit.

INTEGRATION OF THE EXPONENTIAL FUNCTION e^x

Since e^x differentiates to itself, we can now formally state the reverse result that e^x also integrates to itself:

$$\int e^x dx = e^x + c$$

$$\int e^{kx} dx = \frac{e^{kx}}{k} + c$$

where k and c are constants.

INTEGRATION AS THE AREA UNDER A CURVE AND THE LIMIT OF A SUM

Consider a completely general function

$$y = f(x)$$

and the area underneath it (down to the x axis) between the lines $x = a$ and $x = b$ (see Figure 6.7).

Consider a narrow strip parallel to the y axis of (general) height y and (very small) width δx. If the strip is considered as a rectangle,

$$\text{Area of strip} = y \cdot \delta x$$

since δx is very small.

We cannot, of course, consider the whole area A underneath the curve as a trapezium, because $y = f(x)$ is a continuous function curving all the time, but we can consider A as made up of a huge number of narrow strips like the one shown; hence, for all strips,

$$A = \text{sum}\,(y \cdot \delta x)$$

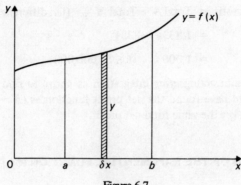

Figure 6.7

In the discrete case we denote this sum Σ, but in this the continuous case we have a sum of infinitely many strip areas, each of infinitely narrow width. This process of δx tending to zero as the number of strips tends to infinity is the *limit* of the sum of the areas, and we call it the *integral*. This is similar to the idea of a limit in differentiation. The sum in the limit is given the symbol $\int dx$, where $\int$ is the Old English S to indicate sum. Thus

$$A = \lim_{\delta x \to 0} (\text{sum } y \cdot \delta x) \text{ between } a \text{ and } b$$

i.e. $A = \int_a^b y\,dx$

Thus, substituting the function $y = f(x)$ and integrating it according to the rule, then subtracting the evaluation at a from the one at b, gives the required area under the curve.

EXERCISES

(1) Find the (indefinite) integrals of the following functions:

(a) $3x^2 + x^{1/2}$.

(b) $2x^4 + 7x - 4$.

(c) $x^3 + \dfrac{2}{x^3} - \dfrac{5}{x^2} + 1$.

(d) $\dfrac{3}{x}$.

(2) Evaluate the following definite integrals:

(a) $\int_{-2}^{2} (4-x^2)dx$.

(b) $\int_{1}^{2} (x^3 + 1)dx$.

(c) $\int_{1}^{8} x^{2/3} dx$.

(d) $\int_{0}^{e-1} \dfrac{dx}{1+x}$.

Hint: Remember that e is just the constant 2.7183 approx.)

(3) Sketch the curve

$$y = 2x^2 - 5x + 2$$

showing clearly where it cuts the co-ordinate axes. (Hint: Calculate max./min. for y to help to sketch the curve).
Find:

(a) the area contained between the curve and the two axes.
(b) the area between the curve and the x axis.

(4) Calculate the area enclosed between the line

$$y = 3x$$

and the curve

$$y = x^2$$

(5) Integrate

$$\frac{dy}{dx} = x^2 - x + 1$$

given that $x = 2$ when $y = 3$.

(6) If

$$f(x) = kx^3(1-x)$$

is known to be a probability density function, for $0 \leqslant x \leqslant 1$, integrate $f(x)$ to find the value of the constant k.

Find the maximum value of $f(x)$, and sketch the graph of the function. (Note: The maximum value is the mode.)

Calculate the probability that x has a value between 0.5 and 0.6.

Find the distribution function

$$F(b) = \int_0^b f(x)dx$$

and hence find the probability that x has a value between 0 and 0.8.

(7) A probability density function is given by

$$p(x) = \tfrac{1}{2}\left(1 - \frac{x}{4}\right) \quad \text{for } 0 \leqslant x \leqslant 4$$

$$= 0 \quad \text{elsewhere}$$

Check that the area under the function for its defined range is unity.
Find:

(a) the distribution function

$$F(X) = \int_0^X p(x)dx$$

(b) the median value (i.e. the x value having half of the area below it and half above it)

(c) the mean value $E(x)$, given the formula for the continuous case of

$$E(x) = \int_0^4 xp(x)dx$$

Compare the discrete formula

$$E(x) = \frac{\Sigma xf}{\Sigma f}$$

Sketch the p.d.f. and the distribution function.

(8) The marginal cost (in £) of producing x thousand Magno batteries is given by the function

$$30(10 - x^{1/2})$$

Find the total cost of production in terms of x.

If production increases from 4,000 to 9,000, what is the total increase in production cost?

(9) Given that marginal revenue is

$$\frac{dR}{dx} = x^2 - 7x + 10$$

where x is the quantity sold, find the revenue function R. (When $x = 0$, assume that $R = 0$.)

Find the average revenue function R/x.

(10) The relationship between income x, consumption C and propensity to save S is given by

$$x = C + S$$

If the marginal propensity to save is known to be

$$\frac{dS}{dx} = a\left(1 - \frac{1}{x^{1/2}}\right)$$

where a is a constant, find the consumption function in terms of x given that consumption is $20a$ when income is 0. (All money units are in £'000,000.)

(11) The efficiency of a worker over his 8 hour shift is assumed to vary continuously at a rate (of change of percentage efficiency w.r.t. time t from starting, in hours) of

$$\tfrac{3}{5}(11 - 4t)$$

If it is known that he commences the shift working at 90% efficiency, find the efficiency function in terms of t.

How efficient is he

(a) after 3 hours?
(b) at the end of the shift?

Sketch the efficiency function.

7 / Probability

Probability is the topic that, in my experience, causes more anguish to students than any other. It seems to start out ludicrously simple, but suddenly the student finds his head spinning, while logical statements couched in clinical language attack the more cobweb-strewn corners of his brain. It is also a topic that gives some lecturers a lot of trouble! Consequently, try to follow each step through carefully, don't overcomplicate things in your mind, and don't scorn what appears to be some trivially obvious statement.

We can do no better than start with the classical definition of the **probability** of a given event as *the ratio of the number of favourable outcomes of the event to the total number of possible outcomes.* This assumes that all outcomes are equally likely to happen; for example,

$$\text{Probability of throwing a five on a fair die} = \tfrac{1}{6}$$

As a convenient notation we write

$$p(\text{five on die}) = \tfrac{1}{6}$$

Similarly,

$$p(\text{draw a four from a pack of cards}) = \tfrac{4}{52} = \tfrac{1}{13}$$

You will find that many of our examples use games of chance and their equipment. (By the way, notice the singular of 'dice'.) This is no coincidence, as historically they gave the incentive to investigate probability, and they still provide useful illustrations.

Two criticisms of the definition, mentioned in passing, are:

(1) What if all outcomes are *not* equally likely (e.g. the probability of a man's dying in n years' time?
(2) What if the number of possible outcomes is infinite (e.g. the time until an event happens)?

A more refined version of our definition is that, as the event is repeated a very large number of times, the ratio approaches a limiting value. Think of tossing a fair coin a million times; you would expect close to half a million heads.

We shall find it very useful immediately to introduce a scale with which to measure the probability of an event. If an event *cannot* happen, there are no favourable outcomes, so our ratio is 0; that is,

$$p(\text{impossible event}) = 0$$

e.g. $p(\text{your swimming the Atlantic Ocean}) = 0$

If an event *must* happen, all the outcomes are favourable, so our ratio is 1; that is,

$$p(\text{certain event}) = 1$$

You may like to suggest some certain events as examples. Most probabilities, of course, lie between these limits of 0 and 1.

Extending our ideas to two events, we say that they are *mutually exclusive* if the occurrence of one means that the other cannot occur; for example, throwing a five means that we cannot at the same time throw a two. The probability of the occurence of one or the other is the sum of their separate probabilities; for example,

$$p(\text{five or six on a die}) = \tfrac{1}{6} + \tfrac{1}{6} = \tfrac{1}{3}$$

Note that 'or' here is used to mean that we don't really mind which. If we use E to denote the particular event, the complementary event (i.e. the particular event does *not* occur) is usually denoted by $\bar{E}$; hence,

$$p(E \text{ or } \bar{E}) = p(E) + p(\bar{E})$$

because they are mutually exclusive. However, one or the other must happen (like the six possible scores on a die, which form a mutually exhaustive set), so

$$p(E) + p(\bar{E}) = 1$$
$$p(\bar{E}) = 1 - p(E)$$

This may appear to be less than world shattering, but it is frequently an elegant way to solve a problem. What we require may be complicated whereas the complementary event is not, so an apparently roundabout method may yield a quick answer.

Two events are said to be *independent* if the occurrence of one does not affect the occurrence of the other; for example, the facts that you were born in September and have brown eyes are unliked to be related. A fair die has no memory; and if you roll a six, there is just the usual probability $\tfrac{1}{6}$ of getting a six on the next roll. If you asked the probability of getting two sixes running before throwing the first, this would be a different event. Incidentally, you may suddenly start to notice the number of people who do not understand the idea of independence. Look out for talk of 'the law of averages', which supposedly proves that a soccer team on a streak of bad luck(!) is more likely to win the next game.

We shall now consider the probability of one event's following another. What is the probability of throwing two heads running on a fair coin, for example? Since each outcome on the first toss can be followed by either of the outcomes on the second, we can show what is happening clearly with a *tree diagram* (see Figure 7.1). This shows that

$$p(2 \text{ heads}) = \tfrac{1}{4} = \tfrac{1}{2} \cdot \tfrac{1}{2}$$
$$= p(\text{first head}) \cdot p(\text{second head})$$

Extend the tree yourself to a third toss of the coin, to show that

$$p(3 \text{ heads}) = \tfrac{1}{8} = (\tfrac{1}{2})^3$$

If events are independent, we multiply their probabilities to find the probability of getting one *and* (i.e. with) the other. Intuitively, the joint occurrence is getting less likely. Note that 'and' here is used to mean that the events are strictly required together.

We have already seen that, if events A and B are mutually exclusive, we add their probabilities to find the probability of one *or* the other's occurring.

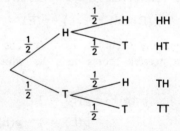

Figure 7.1

Example 1

A bag contains 5 white balls and 7 black balls. If we draw out 2 (without replacement), what is the probability that we have a black and a white?

First, we must realise that we are equally satisfied by the mutually exclusive events 'black then white' or 'white then black', denoted BW and WB respectively; hence,

$$p(\text{B and W, any order}) = p(\text{BW}) + p(\text{WB})$$

Because there is no replacement, the probability of the second ball's being a particular colour is changed. We say here that we have a **conditional probability**, because an event is **dependent** on some condition already imposed. (For example,

$$p\begin{pmatrix}\text{you will live to be 70} \\ \text{when born}\end{pmatrix} \neq p\begin{pmatrix}\text{you will live to be 70} \\ \text{when you have reached 30}\end{pmatrix}$$

because your context will have changed.) In this case there are only eleven balls left for the second draw, not twelve; hence,

$$p(BW) = \tfrac{7}{12} \cdot \tfrac{5}{11} = \tfrac{35}{132}$$
$$p(WB) = \tfrac{5}{12} \cdot \tfrac{7}{11} = \tfrac{35}{132}$$
$$\therefore p(B \text{ and } W) = \tfrac{35}{132} + \tfrac{35}{132} = \tfrac{35}{66}$$

Example 2

A firm owns two machines, which have different probabilities of running throughout the week without breaking down. The newer one has probability of 0.8 and the older one a probability of 0.75. What is the probability in a particular week that:

(a) both run without breakdown?
(b) at least one breaks down?

Solution:

(a) p (both run) = 0.8 × 0.75 = 0.6
(b) p (at least one breaks down) = $1 - p$(neither breaks down) = $1 - 0.6 = 0.4$

Notice there are three possibilities included in p (at least one breaks down), but our method shortcircuits the need to work them out and add them up.

Example 3

What is the probability of throwing a total of 6 on 2 dice?

We could draw a tree diagram with 6 branches leading each to 6 more (giving a total of 36 outcomes), but often it is easier systematically to list the possibilities of that total. Since we have 2 different dice, the possibilities are

1 and 5 or 5 and 1
2 and 4 or 4 and 2
3 and 3

$\therefore p(\text{total 6 on 2 dice}) = \frac{5}{36}$

Example 4

Among 80 students in the first year of a course, 48 are men; 22 of the men have 'O' level Mathematics, and 35 students altogether have 'O' level Mathematics. One of the students is chosen at random (everyone has an equal chance of selection) to be the representative of the year. What is the probability that:

(a) the representative will be a woman without 'O' level Maths?
(b) the representative will be a woman, given that he/she has 'O' level Maths?
(c) the representative will have 'O' level Maths, given that representatives are men?

This example is to demonstrate further the idea of conditionality. At first sight it appears complicated, but analysing the situation with a table quickly clarifies it. First we fill in what the question tells us (see Table 7.1), and then we complete the table by simple arithmetic (see Table 7.2).

Table 7.1

	With 'O' level Maths	Without 'O' level Maths	Total
Men	22		48
Women			
Total	35		80

Table 7.2

	With 'O' level Maths	Without 'O' level Maths	Total
Men	22	26	48
Women	13	19	32
Total	35	45	80

(a) There are 19 women without 'O' level Maths out of a total possible of 80 (no condition has been imposed or is presumed known), so

$$p_1 = \tfrac{19}{80}$$

(b) There are 35 with 'O' level Maths (the condition presumed), of whom 13 are women, so

$$p_2 = \tfrac{13}{35}$$

(c) There are 48 men, of whom 22 have 'O' level Maths, so

$$p_3 = \tfrac{22}{48}$$

Notice the notation employed here with the probability answers, in order to avoid a boring repetition of words.

EXERCISES

(1) Find the probability that:
   (a) an even number appears in one throw of a fair die.
   (b) a picture card appears in one draw from a pack of cards.
   (c) both the events (a) and (b) above happen together.
   (d) 2 heads and 1 tail turn up when a fair coin is thrown 3 times.

(2) What is the probability of drawing the ace, king and queen of hearts in a draw of 3 cards from a pack?

(3) Cards are drawn without replacement from a full pack. What is the probability that 5 cards will precede the first ace?

(4) A manufactured item is composed of 3 components, which have probabilities of being defective of 1%, 4% and 2% respectively. What is the probability that:
   (a) a particular item has all components defective?
   (b) a particular item contains at least 1 defective component?

(5) If 2 fair dice are thrown, find the probability that the combined score is greater than or equal to 9.

(6) A husband and wife are judged by an insurance firm to have the probability of each living for another 25 years of $\tfrac{2}{3}$ and $\tfrac{4}{5}$ respectively. Find the probability that:

(a) both are alive after 25 years.
(b) only the man is alive after 25 years.
(c) at least one is alive after 25 years.

(7) If 3 dice are thrown, what is the probability that they all show different numbers?

(8) If 2 components are known to be faulty in a box containing 10, and 2 are selected at random, find the probability that:

(a) both are faulty.
(b) one is faulty.
(c) neither is faulty.

(9) If 3 marksmen fire at a target, with respective probabilities of hitting it of $\frac{1}{2}$, $\frac{3}{4}$ and $\frac{2}{3}$, find the probability that:

(a) only 1 hits the target.
(b) only 2 hit the target.

(10) There are 100 young people applying for a job; 35 of them have previously held a job, and 25 of these have qualifications, out of the 40 in all who have some qualifications. What is the probability that an applicant:

(a) has neither held a job previously nor has qualifications?
(b) has no qualifications?
(c) has previously held a job, given that he has qualifications?
(d) has qualifications, given that he has not held a previous job?

8. Sets

A *set* is basically a collection of things. We usually give each set a capital letter to name it and describe its members (*called elements*) in a curly bracket; for example, the set of vowels is written

$$V = \{a, e, i, o, u\}$$

The language of sets is useful in probability theory, and the pattern of the theory of sets is like that of logic, which has important applications in the design of logical circuits for computers.

Consider the set of possible outcomes when we throw a coin twice. We can write, using H for a head and T for a tail,

$$A = \{HH, HT, TH, TT\}$$

This is, of course, a finite set. We can also have an infinite set to describe an inequality, such as we investigated in Chapter 2:

$$S = \{(x, y) \mid 2x + 3y \geq 6\}$$

This means: all the points (x, y), such that the inequality

$$2x + 3y \geq 6$$

is satisfied. It therefore describes the region left unshaded (and the boundary) in Figure 2.4.

Consider another example:

$$B = \{x \text{ an integer} \mid 0 < x \leq 8\}$$

Here, B is the set of numbers $\{1, 2, 3, 4, 5, 6, 7, 8\}$.

A *null set* or *empty set* is a set that has no members. We write

$$\phi = \{\quad\}$$

or just refer to it as ϕ.

A *subset* is a set that is completely contained in another set. For example, considering the possible outcomes of throwing a die,

$$C = \{1, 3, 5\} \quad \text{(the odd throws)}$$
$$D = \{1, 2, 3, 4, 5, 6\}$$

then C is a subset of D; we write

$$C \subset D \quad \text{(i.e. } C \text{ is contained in } D)$$
$$D \supset C \quad \text{(i.e. } D \text{ contains } C)$$

Furthermore, given

$$E = \{1, 2, 3, 4, 5, 6, 7, 8\}$$

we can extend the connection further and say

$$C \subset D \subset E$$
$$\text{and} \quad C \subset E \quad \text{(of course)}$$

The *union* of two sets A and B is the set of elements that are in *either A or B or both*. We do not count twice any element that is in both. We write the union $A \cup B$. For example,

$$\text{if} \quad A = \{\text{cup, saucer, plate}\}$$
$$\text{and} \quad B = \{\text{cup, saucer, knife, fork}\}$$
$$\text{then} \quad A \cup B = \{\text{cup, saucer, plate, knife, fork}\}$$

The *intersection* of two sets A and B is the set of elements that are in *both A and B* (i.e. strictly common to both). We write the intersection $A \cap B$. In the previous example

$$A \cap B = \{\text{cup, saucer}\}$$

We can relate union and intersection to probability theory here by noting that, if A and B represent two events,

$A \cup B$ means that either A has happened or B has happened or both A and B have happened (provided that they are not mutually exclusive)

$A \cap B$ means that both A and B have happened

The *universal set* (i.e. sample space) U is the set of all the possible elements in the context that we are considering. Thus, if we were considering only the letters of the English alphabet, this would be U.

The *complement* of a set A is the set of all the elements in U that are not in A. We shall use the notation A' for the complement. For example,

$$\text{if} \quad V = \{\text{vowels}\}$$
$$\text{and} \quad U = \{\text{letters in alphabet}\}$$
$$\text{then} \quad V' = \{\text{consonants}\}$$

We shall now show these ideas pictorially using a *Venn diagram*. This is just a rectangle showing the sets as circles, containing their elements, which may or may not overlap.

We sometimes shade sets to indicate the elements within. Notice in Figure 8.1 that

$$A \cup B = \text{Area shaded in any manner}$$

$$A \cap B = \text{Area double shaded}$$

Since $A \cup B$, like all combinations of sets, is itself a set, notice that $(A \cup B)'$ (i.e. the complement of $A \cup B$) is the area within the universe left unshaded.

It can be convenient to indicate the elements in the sets as in Figure 8.2.

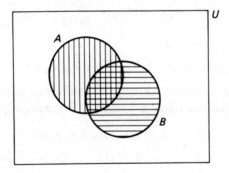

Figure 8.1

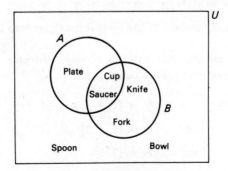

Figure 8.2

Sometimes only the number of elements in each region is written, as in Example 3.

If we use the notation $n(A)$ to mean the number of elements in set A, Figure 8.2 gives

$$n(A) = 3 \quad n(A \cup B) = 5 \quad n(U) = 7$$
$$n(B) = 4 \quad n(A \cap B) = 2 \quad n(\phi) = 0 \quad \text{(by definition)}$$

The Venn diagram shows that

$$n(A \cup B) = n(A) + n(B) - n(A \cap B)$$

since, unless we subtract the intersection, we have counted it twice.

If we think of A and B as events and a probability as our ratio,

$$\text{e.g.} \quad p(A) = \frac{n(A)}{n(U)}$$

then we can see similarly that

$$p(A \cup B) = p(A) + p(B) - p(A \cap B)$$

If the events are mutually exclusive, the intersection is empty, and we have the simpler result obtained in Chapter 7. For example,

if $\quad U = \{1, 2, 3, 4, 5, 6, 7, 8, 9, 10\}$

and $\quad A = \{2, 4, 6, 8, 10\} \quad$ (the even numbers)

and $\quad B = \{3, 6, 9\} \quad$ (multiples of 3)

then $\quad p(A) = \frac{5}{10} \quad p(B) = \frac{3}{10} \quad p(A \cap B) = \frac{1}{10}$

$\therefore p(A \cup B) = \frac{5}{10} + \frac{3}{10} - \frac{1}{10} = \frac{7}{10}$

That is, there are 7 numbers of the 10 that are either even or multiples of 3 or both.

Disjoint sets are sets that have no common elements. They are not necessarily complementary. For example,

if $\quad D = \{1, 2, 3, 4, 5, 6\}$

and $\quad C = \{1, 3, 5\}$

and $\quad F = \{2, 6\}$

then C and F are disjoint.

Example 1
Simplify the expression $(A' \cap B')'$ as much as possible.

We draw a completely general Venn diagram with two sets and attempt to shade well-defined regions, working out in steps from the inside of the expression. From Figure 8.3 we can see that the intersection of A' and B' (i.e. $A' \cap B'$) must be the area shaded both ways (shown separately in Figure 8.4). Therefore its complement must be the unshaded area in Figure 8.4, which is clearly $A \cup B$; that is

$$(A' \cap B')' = A \cup B$$

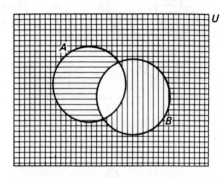

A' shaded ≡
B' shaded |||

Figure 8.3

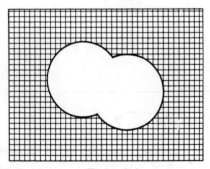

Figure 8.4

Example 2
Prove that

$$A \cap (B \cup C) = (A \cap B) \cup (A \cap C)$$

The left hand side is depicted in Figure 8.5, where the intersection of A and $B \cup C$ is the double-shaded area (shown separately in Figure 8.6).

The right hand side is depicted in Figure 8.7, where the union of $A \cap B$ and $A \cap C$ is the area shaded either way or both (shown separately in Figure 8.8).

Clearly, Figures 8.6 and 8.8 depict the same area, so the equation is proved.

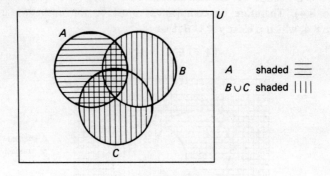

Figure 8.5

Figure 8.6

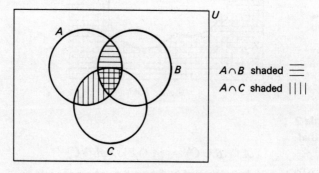

Figure 8.7

Figure 8.8

Example 3

There are 94 employees in a firm, of whom 57 are men. Of the men 35 are married and 32 are over 25 years old, but only 20 married men are over 25 years old. There are altogether 60 employees over 25 years old; and there are 60 married people, of whom 39 are over 25 years old. How many unmarried women under 25 years old work at the firm?

We first set up a Venn diagram with three overlapping circles to represent the three sets of employees who are married, who are men and who are over 25. We look in the information for something very well defined and find a region in all three sets (i.e. the intersection of all three), which is married men over 25. This region is shown in Figure 8.9, where set totals of members have also been inserted in brackets.

The crucial point is to realise that saying 35 men are married (i.e. lie in the intersection of these two sets) means that some married men are over 25 and some are not. Using our most exclusive triple intersection, we can subtract 20 from 35 to discover that 15 married men are *not* over 25. Similarly, we find that 12 men are over 25 but *not* married and 19 married people over 25 are *not* men (see Figure 8.10).

It is then an easy task to complete the three set totals; for example, 6 married people are *not* men *nor* over 25 (i.e. married women under 25) (see Figure 8.11).

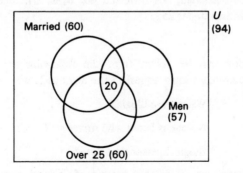

Figure 8.9

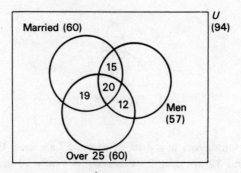

Figure 8.10

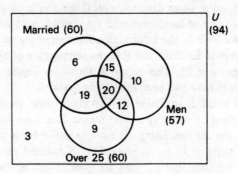

Figure 8.11

Adding all the separate region figures together we have 91 employees. The universal set is 94 employees, so just 3 must lie outside all three sets (i.e. are *not* married, *not* men and *not* over 25); that is, there are 3 unmarried women under 25.

Example 4

What conclusion can be drawn from the following statements? (The context is deliberately bizarre to reflect the work of Lewis Carroll.)

> Waiters are illogical
>
> No one is hated who drives well
>
> Illogical persons are hated

The first statement means that the set of illogical persons I contains all waiters W (it may also contain others, but no waiters are outside it).

This can be expressed either mathematically, as

$$W \subset I$$

(using the obvious symbols for each set), or pictorially with circles, as in Figure 8.12.

Similarly, the persons who drive well D are contained in the set of people who are not hated H'; that is,

$$D \subset H' \quad \text{(also } H \subset D') \quad \text{(see Figure 8.13)}$$

Finally, all illogical persons I must be within the hated set H (with possibly others); that is,

$$I \subset H \quad \text{(see Figure 8.14)}$$

Putting the diagrams together, and noting that, because they are complements, it is easy to split the universe into hated and not hated, we get Figure 8.15. From this it is clear that

$$W \subset I \subset H$$

$$\therefore W \subset H \quad \text{but} \quad D \subset H'$$

Therefore, W does not intersect D, or

$$W \cap D = \phi$$

or, as can be seen from the diagram, the sets of waiters and persons who drive well have no common elements; that is,

Waiters do not drive well

If you were able to see that $H \subset D'$, you could perhaps give a direct alternative solution.

Figure 8.12

Figure 8.13

Figure 8.14

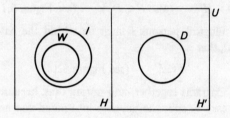

Figure 8.15

EXERCISES

(1) What is ϕ'?

(2) Is $\{\phi\}$ empty?

(3) If $A \subset B$, what is:

    (a) $A \cup B$?
    (b) $A \cap B$?

(4) What is:

    (a) $A \cup A'$?
    (b) $A \cap A$?

(5) Simplify the following expressions as much as possible, using a Venn diagram:

    (a) $A \cap (A' \cup B)$.
    (b) $(A \cap B) \cup (A \cap B')$.

(6) Prove that

$$A \cup (B \cap C) = (A \cup B) \cap (A \cup C)$$

(7) Use set notation to describe the shaded area in the diagram (Figure 8.16).

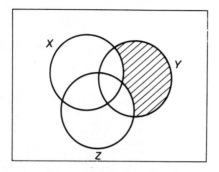

Figure 8.16

(8) The sets L, M and N in a universal set U consisting of the first 10 lower-case letters of the alphabet are defined as:

$$L = \{a, b, c\}$$
$$M = \{b, c, d, e\}$$
$$N = \{a, d, e, f\}$$

Find $M \cap N$, $L \cup N$ and $L \cap M \cap N$.
Find also L' and $(L \cup M \cup N)'$.
How can L, M and N be combined to produce the set $\{a, d, e\}$?

(9) On a particular day the numbers of workers in a small cement-block factory who worked the early, late and night shifts were 20, 45 and 30 respectively. However, 12 workers 'jagged on' (i.e. worked on the late as well as the early shift). If no worker did both early and night shifts, how many worked both late and night shifts, given that in all 75 different workers clocked in on that day?

(10) Use set notation to show whether the following argument is valid:

All Spaniards are short

No short person is haughty

Therefore, no haughty person is Spanish

(11) What conclusion can be drawn from the following statements?

Persons who are respectful to authority always play cricket

All Australians are dogged in adversity

You cannot play cricket unless you drink beer

Persons who are dogged in adversity are never disrespectful to authority

9 Introduction to Statistics

DEFINITIONS

You will quite possibly have approached this chapter with the old quotation about 'lies, damned lies and statistics' symbolising for you an attitude of mixed fear and contempt towards the subject. Neither is at all justified. As you advance in your other studies, the importance of statistics will surprise you, but you must guard against errors both accidental and intended by unscrupulous people. It is not true that 'anything can be proved with statistics', but some people do try.

Science generally is increasingly becoming aware that phenomena cannot always be approached in a cut and dried manner. Very often, we must resort to a probability statement about some activity, such as the control of the quality of a product or the prediction of a human behaviour trait. Forecasting the future may be difficult and sometimes amusing, but without forecasts no plans or decisions could be made at all. Every subject that you study in business studies needs forecasts, and statistics is the tool used.

We are at first concerned in this book with collecting information (called *data*), organising and summarising it into a comprehensible form, presenting it and then analysing it to some extent. This is *descriptive statistics*. Later will come testing and the drawing of valid conclusions in order to make rational decisions. A figure obtained from data is often described as a *statistic*. *Raw data* is collected information that has not yet been processed.

We shall deal with two types of variable. A *discrete variable* can only take whole number values (e.g. the number of students taking an examination). A *continuous variable* can take any value and presumes that measurement is possible to any (required) accuracy (e.g. the height of a student). Often, we only measure to a certain accuracy, and a theoretically continuous curve is really a series of blocks.

A *population* is the whole group of things under consideration, like the universal set in set theory, and does not have to be people. It can be small, large or infinite.

A *sample* is a selection made from the population. Exactly how samples are selected (e.g. to be representative or random) is a topic in its own right and does not concern us yet. *Random selection* implies that every member of the population has an equal chance of selection. A conclusion (called a statistical *inference*) derived from analysing data is couched in probability language and is usually dependent for validity on the random selection of the sample.

THE FREQUENCY DISTRIBUTION

Suppose the marks obtained by 50 students in a mathematics examination are as given in Table 9.1. As presented, these raw figures are complete but suggest no pattern, so we shall organise them into **classes**. Information is obviously lost, but we hope that there will be a gain outweighing the loss.

Table 9.1 *Students' marks: raw data*

66	58	57	59	60
64	53	57	62	59
42	34	41	71	43
53	(24)	58	45	48
50	57	44	33	30
41	64	43	54	59
51	49	58	75	(78)
42	61	55	54	61
45	56	45	66	38
39	43	42	29	50

Generally, we first look for the smallest and the largest figures. The difference between them is the *range*. Here

$$\text{Range} = 78 - 24 = 54$$

We now try to divide the range into a convenient number of classes, all with the same width or *class interval*. The total frequency is also important, because we do not want many classes with small *class frequency* (i.e. the number in each class), nor do we want few classes with large class frequency. Here, let us choose 6 classes, starting with limits 20 and 29 and going up to limits 70 and 79. We now go through the data, ticking them off into classes, and construct a *frequency distribution* (see Table 9.2).

A *class boundary* is considered to be halfway between the upper and lower limits of adjacent classes. This is to suggest continuity, even though

Table 9.2 *Students' marks: frequency distribution*

Mark, x					Frequency, f
20–29	11				2
30–39	⊬⊬⊬				5
40–49	⊬⊬⊬	⊬⊬⊬	1111		14
50–59	⊬⊬⊬	⊬⊬⊬	⊬⊬⊬	111	18
60–69	⊬⊬⊬	111			8
70–79	111				3
					50

only whole marks are possible. Therefore, class 20–29 has boundaries at 19.5 and 29.5 and interval of 10. The *class mark* is taken as the midpoint (e.g. 24.5 in class 20–29); and now that individual figures are lost, we assume that all the members of a class have its class mark (e.g. 18 values of 54.5 in class 50–59).

Note: When groups or classes occur with frequency of zero, it is often convenient to combine with another group whose frequency is not zero.

GRAPHICAL PRESENTATION OF DATA

Before we go further with our example, let us consider various ways in which you meet data on television, in newspapers, and so on.

A *line graph* is shown in Figure 9.1. The x axis is often time, and we then have a graph of a *timeseries*. Whenever possible, the origin must be shown.

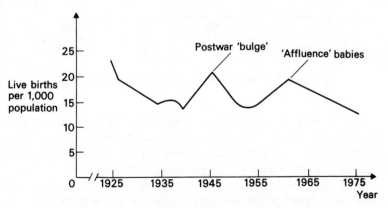

Figure 9.1 Birth rate for England and Wales, 1925–75

A *bar chart* is shown in Figure 9.2. Notice that the size of the figures has caused the origin to be not shown correctly. The effect is usually a distorted and therefore more alarming comparison. More blatant tricks are to magnify or shrink axes or not even to label them.

A *histogram* is an upright version of a bar chart and we can draw one for our example of the students' marks (see Figure 9.3).

A *frequency polygon* does the same job as a histogram, but each frequency is shown as a point corresponding to the class mark of each class instead of as a block (see Figure 9.4). It is sometimes superimposed on to the histogram. Sometimes the frequencies are converted into percentages of the total frequency. At each end the polygon can be completed by joining to the midpoint on the x axis of what would have been the next group below or above.

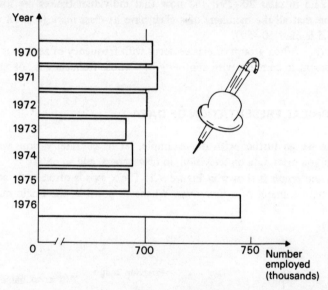

Figure 9.2 Growth of the Civil Service in the 1970s

Another form of horizontal or upright bar chart has no particular name but shows how a complete entity is divided into sections or components. Figure 9.5 is such a chart, showing public spending. Attempts to show this as a pile of money (i.e. by volume) lead to distortion.

A *pie chart* uses the same idea of divisions but is shown on a circle, sometimes referred to by colourful names like 'the national cake' (see Figure 9.6). Total magnitude cannot, of course, be conveyed.

Introduction to Statistics 119

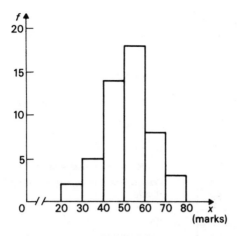

Figure 9.3 Students' marks: histogram

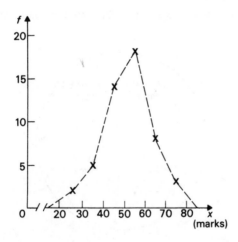

Figure 9.4 Students' marks: frequency polygon

A *pictogram* is really a bar chart that shows the size of each bar with pictures of the object represented. Like bar charts, pictograms often compare two different countries or factories within the chart (see Figure 9.7). An obvious criticism is that fractions of a figure representing many items do not indicate the size. Often the totals are not given at all, but as a visual comparison the pictogram has merit.

In our example of student marks we may be more interested in the number of students getting less than a certain mark (e.g. a pass mark of

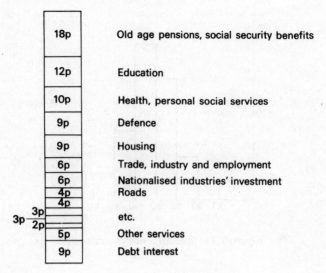

Figure 9.5 How each £1 of public spending was spent in 1978

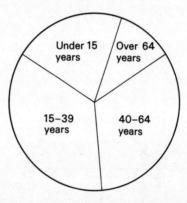

Figure 9.6 Population of England and Wales, 1971 (total 50 million)

40 or 50) than we are in a particular class interval. We construct a *cumulative frequency distribution* or *ogive* by adding the figures up as we move higher through the marks (giving a 'less than' ogive). Thus, there are (2 + 5) marks less than 39.5 and (2 + 5 + 14) marks less than 49.5, and so on, until finally all the marks are less than 79.5 (see Table 9.3).

Until we reach the upper boundary of each class interval, we cannot assume that all the marks have been included; so when plotting an ogive, we plot against the *upper* boundary. This is most important. We earlier

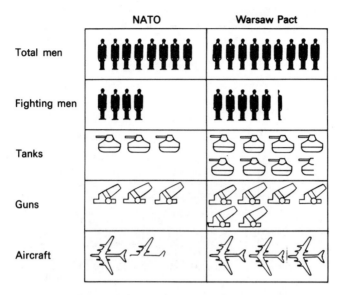

Figure 9.7 Military balance of NATO and Warsaw Pact, 1975

Table 9.3 *Students' marks: cumulative frequency distribution*

Mark	Class mark	f	Cumulative 'less than'
20–29	24.5	2	2
30–39	34.5	5	7
40–49	44.5	14	21
50–59	54.5	18	39
60–69	64.5	8	47
70–79	74.5	3	50
		50	

said that all the marks in a class were assumed, once grouped, to be at the class mark. This is the same as assuming that they spread perfectly evenly throughout the class; for example, 5 marks in class 30–39 plot as in Figure 9.8. The broken S shape of an ogive is typical and gives the ogive its name (see Figure 9.9).

We could alternatively construct a 'more than' cumulative frequency distribution or ogive, but in practice we rarely need both.

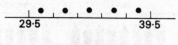

Figure 9.8

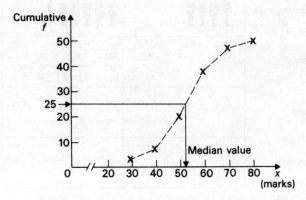

Figure 9.9 Students' marks: ogive ('less than')

MEDIAN AND MODE

We shall slightly anticipate the next chapter by noting two things. First, the histogram shows us the *modal class* (i.e. the most frequently-occurring value). Secondly, the ogive allows us to estimate the *median value* (i.e. the value that has half of the total frequency above it and half below it). In this example, with 50 students, we have 25 above and 25 below at the 25th frequency. Drawing across from $f = 25$ to the ogive then down to the x axis, as shown in Figure 9.9, we find the median. It is typical of the examination marks in that it is central in terms of total frequency, and this very important idea of typicality is developed further in the next chapter.

EXERCISES

(1) The number of telephone calls received at an exchange in 100 successive intervals, each of 30 seconds duration, is given in the table. Form a frequency table and a cumulative frequency table of the data.

Number of calls

6	3	4	6	5	1	2	2	4	4
4	4	6	4	6	5	7	1	1	0
0	3	3	5	5	4	6	4	6	2
5	3	3	6	4	4	2	3	3	4
2	1	6	6	3	3	4	4	5	3
5	2	7	2	6	5	4	3	3	7
4	4	7	3	5	4	5	3	1	6
1	4	6	3	3	0	5	1	6	5
6	3	1	4	1	5	6	5	6	1
6	7	5	3	6	1	3	7	2	5

(2) A mail order firm takes a sample of 50 of its orders, which are given in the table. Construct a frequency distribution and a cumulative frequency distribution.

Draw the histogram, frequency polygon and both ogives.
What percentage of orders is between £10 and £39?
Estimate the median order value.

Value of order (£)

36	11	16	21	29	39	16	29	37	8
48	33	12	22	16	3	39	13	25	9
11	86	47	26	6	58	38	8	18	28
7	29	19	34	44	25	13	31	9	22
23	32	41	13	52	4	24	27	15	17

(3) The table gives the frequency distribution of marks obtained in an examination. Draw the histogram, the frequency polygon and the cumulative frequency ('less than') distribution for the data.

Find the median value.

Mark	Number
30–39	1
40–49	4
50–59	8
60–69	20
70–79	35
80–89	26
90–99	6

(4) Cut out examples of data presentation that you find in newspapers, and either bring them in to class or keep them with your notes.

10 Measures of Location and Dispersion

SUMMATION NOTATION

It will be helpful to you, and extremely useful in this and further chapters, if we develop a little further our notation, for summation, of Σ.

We have met the use of f to denote frequency, but how can we denote the frequency of a particular class or group? If we give the classes numbers 1, 2, 3, and so on, we can refer to their frequencies as f_1, f_2, f_3, and so on and to the class mark of each group as x_1, x_2, x_3, and so on. This subscript notation would give, in our student mark example, a group 3 of 40–49 with $x_3 = 44.5$ and $f_3 = 14$.

The summation notation enables us to write a shorthand form such as

$$\sum_{i=1}^{6} x_i = x_1 + x_2 + x_3 + x_4 + x_5 + x_6$$

meaning 'add all the values of x_i, starting from x_1 (when $i = 1$) and finishing at x_6 (when $i = 6$)'. If the number of values is understood, we sometimes write Σx instead of $\sum_{i=1}^{6} x_i$. Similarly, we can write

$$\sum_{i=1}^{6} x_i f_i = x_1 f_1 + x_2 f_2 + x_3 f_3 + x_4 f_4 + x_5 f_5 + x_6 f_6$$

or just Σxf. For example, using the frequencies for the set of numbers

$$\{1, 2, 3, 4, 5, 6\} \qquad \text{(Set 1)}$$

given in Table 10.1, we get

$$\Sigma x = \sum_{i=1}^{6} x = 1 + 2 + 3 + 4 + 5 + 6$$

$$= 21$$

$$\Sigma xf = \sum_{i=1}^{6} x_i f_i = 1(2) + 2(5) + 3(14) + 4(18) + 5(8) + 6(3)$$

$$= 2 + 10 + 42 + 72 + 40 + 18$$

$$= 184$$

Similarly,

$$\Sigma f = \sum_{i=1}^{6} f_i = f_1 + f_2 + f_3 + f_4 + f_5 + f_6$$

$$= 2 + 5 + 14 + 18 + 8 + 3$$

$$= 50 \quad \text{(i.e. total frequency)}$$

Can you see what

$$\frac{\Sigma xf}{\Sigma f} \left(= \frac{184}{50} = 3.68 \right)$$

may represent? If not, we shall return to it shortly.

Table 10.1 Set 1: frequency distribution

i	1	2	3	4	5	6
x_i	1	2	3	4	5	6
f_i	2	5	14	18	8	3

MEASURES OF LOCATION

One of the most often heard but most commonly misused words in the English language is *average*. You hear every day talk of the 'average wage' of £x, the 'average family' with its 2.4 children, and so on. Sometimes different sides to a dispute produce different average numbers to describe what should be the same object, causing at best confusion and at worst bitter disagreement. How can this be?

There are in fact several types of average. Each tries in its own way to be typical of, or to represent, a set of data. They try to locate where the data can be typically thought of as being and are also known as *measures of central tendency*.

Consider the following set of numbers:

$$\{5, 4, 7, 3, 10, 6, 7, 8, 7, 3\} \quad \text{(Set 2)}$$

It is convenient to rewrite them in order of size; that is,

$$\{3, 3, 4, 5, 6, 7, 7, 7, 8, 10\}$$

We have already mentioned (in Chapter 9) the *mode* (or **modal group** if we have a group rather than one figure), which is the most fashionable or most frequently occurring number. Here it is 7. If two (or more) numbers occur with the same most-common frequency, the mode is not unique. We use expressions like *bimodal* in such cases.

The *median* or halfway value (halfway in frequency, *not* halfway up the range) has also been mentioned. Here it is between the 5th and 6th numbers, i.e. at

$$\tfrac{1}{2}(6 + 7) = 6.5$$

If the total frequency is odd, say 39, then we take the 20th value, because 19 are below and 19 above. We *must* have the data in ascending order to find the median. The median always exists and is unique.

Both the mode and median are particularly useful for non-quantifiable objects, but the most common form of average is the *arithmetic mean*. All the numbers are added together and divided by how many numbers there are. If the numbers are $x_1, x_2, \ldots x_n$, we use the symbol $\bar{x}$, hence,

$$\bar{x} = \frac{\sum_{i=1}^{n} x_i}{n}$$

In our simple example (set 2),

$$\bar{x} = \frac{3 + 3 + 4 + 5 + 6 + 7 + 7 + 7 + 8 + 10}{10} = \tfrac{60}{10} = 6$$

The arithmetic mean is typical of the data because it 'balances' the numbers; it is the result of evening out the higher and lower numbers (here, as if there were 10 equal numbers). It is predominant, because it uses all the values, always exists and is unique. It is readily susceptible to further mathematical manipulation. It is reliable in that it varies less than other averages when repeated samples are taken (e.g. heights of ten men at a time, taken from a population) and is extremely important in our further work in statistical inference.

One disadvantage is that an extreme figure in the data can distort the mean. If you consider as an example the earnings of all the people in a firm, the mean earnings may be untypical because of a few directors with high salaries. The modal earnings may be untypical because they belong to a most-common but still minority group on low wages. The median is probably best here.

Each situation requires its own best average. There are other averages besides these three, such as the *geometric mean*, used for growth rates, and others that you could investigate (e.g. the *harmonic mean* and the *golden mean*).

In practice, the formula that we have already for the (arithmetic) mean needs extending, because real life figures occur in groups, not individually. Suppose that we group the figures of our simple example (see Table 10.2). The xf column merely saves us the trouble of adding each x value separately. We would never add 12 equal heights of 5 ft 10 in. separately, for example; instead we multiply to get the total contribution of 12 times 70 in. (i.e. 840 in.) from that height category. The end result is, of course, identical, but we have calculated

$$\bar{x} = \frac{\Sigma fx}{\Sigma f}$$

Table 10.2 *Set 2: calculation of weighted arithmetic mean*

x	f	xf
3	2	6
4	1	4
5	1	5
6	1	6
7	3	21
8	1	8
9	0	0
10	1	10
	10	60

$$\bar{x} = \frac{\Sigma xf}{\Sigma f} = \frac{60}{10} = 6$$

This is a *weighted arithmetic mean*. Each value x has been weighted (i.e. multiplied) by its frequency, which is obviously the measure of its importance, before adding the products up and dividing by the total frequency.

Let us finally work through an example to find the arithmetic mean of grouped data. Consider the income figures in Table 10.3. We can rewrite the income column using the class mark as the value x_i for each class. Note that we use common sense in taking £25, £35, and so on instead of £24.995. We have an open-ended group but take the class mark as £65 for consistency (see Table 10.4). Thus, we get a mean weekly income of £46.5.

Table 10.3 *Weekly income: frequency distribution*

Weekly income £	Number of persons
20–29.99	5
30–39.99	18
40–49.99	42
50–59.99	27
60 and above	8
	100

Table 10.4 *Weekly income: calculation of mean*

x	f	xf	u	uf
25	5	125	1	5
35	18	630	2	36
45	42	1,890	3	126
55	27	1,485	4	108
65	8	520	5	40
	100	4,650		315

$$\bar{x} = \frac{\Sigma xf}{\Sigma f}$$

$$= \frac{4,650}{100} = 46.50 \quad \text{(i.e. £46.5)}$$

With widespread use of calculators nowadays the calculation even with large values of x and f is not too daunting, but it is often more convenient to use a simplified scale for x, also called **coding**. The subtraction of a large common number from every x value, say £15 in this case, will not affect the answer if we add it back to the mean at the end. Similarly, the gaps of £10 are all equal, so we can reduce (i.e. shrink) the values by dividing them all by 10 (see Table 10.4); that is,

$$u_i = \frac{x_i - 15}{10}$$

e.g. $$u_3 = \frac{x_3 - 15}{10}$$

$$= \frac{45 - 15}{10} = 3 \quad \text{(i.e. £3)}$$

All this can be remembered with formulae, but it is much better to realise what you are doing and just reverse the process at the end; hence,

$$\bar{u} = \frac{\Sigma uf}{\Sigma f}$$

$$= \frac{315}{100} = 3.15 \quad \text{(i.e. £3.15)}$$

We must magnify back by a factor of 10 and then add on £15, to get

$$\bar{x} = (3.15 \times 10) + 15 = 31.5 + 15 = 46.5 \quad \text{(i.e. £46.5, as before)}$$

We really have two parallel scales, as shown in Figure 10.1.

The median can also be calculated using a formula, but it is again better to just understand what we are doing. In the weekly income example the median is the 50th value (of 100). We can count cumulatively but only get 23 values before reaching the 42 values in the class £40–49.99. We need 27 more values of these 42. Assuming, as we have done before, that they are all spread evenly between £40 and £49.99, we just need to penetrate by the correct proportion of the interval, which is

$$\frac{27}{42} \times 10 = 6.43 \quad \text{(i.e. £6.43)}$$

Thus, with £40 given by the first 23 values and £6.43 given by the next 27,

$$\text{Median} = 40 + 6.43 = 46.43 \quad \text{(i.e. £46.43)}$$

The modal group is £40–49.99. It is possible to calculate an exact figure for the mode by using a formula, which is best understood by referring to the histogram in Figure 10.2:

$$\text{Mode} = L + \left(\frac{a}{a+b}\right) \cdot I$$

$$= 40 + \left(\frac{24}{24+15}\right) \cdot 10$$

$$= 40 + \left(\frac{24}{39}\right) \cdot 10$$

$$= 40 + 6.15$$

$$= 46.15 \quad \text{(i.e. £46.15)}$$

You should be able to see now that, when we calculated $\Sigma xf/\Sigma f = 3.68$ earlier in the chapter, it gave the mean student examination mark. As an exercise complete the decoding process to arrive at the mark.

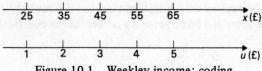

Figure 10.1 Weekley income: coding

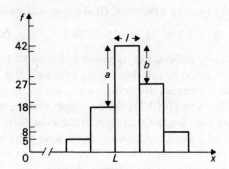

Figure 10.2 Weekly income: calculation of mode

The mean, median and mode will only be the same for a symmetrical distribution.

MEASURES OF DISPERSION

As well as wanting to know a measure of the location of data (i.e. where it is) we also want to know how it is spread out, varies or is dispersed.

Let us consider diagramatically in Figure 10.3 the simple set of numbers that we had previously, namely,

$$\{3, 3, 4, 5, 6, 7, 7, 7, 8, 10\} \qquad \text{(Set 2)}$$

The most obvious *measure of dispersion* is the range, which here is $10 - 3 = 7$. However, the range has a clear disadvantage in its possible distortion by extreme values above or below.

It can be refined, using an extension of our median idea, as the *interquartile range*. The quartiles are values that break the total frequency into groups of 25 per cent:

$$\begin{aligned}
\text{First quartile} &= Q_1 = 25\% \text{ value} \quad \text{(in frequency)} \\
\text{Second quartile} &= Q_2 = 50\% \text{ value} \quad \text{(i.e. median)} \\
\text{Third quartile} &= Q_3 = 75\% \text{ value}
\end{aligned}$$

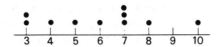

Figure 10.3   Set 2: frequency distribution

Hence,
$$\text{Interquartile range} = Q_3 - Q_1$$

It is thus the range of values occupied by the middle 50 per cent of values. Another measure is the *quartile deviation or semi-interquartile range* (SIQ); that is,

$$\text{Quartile deviation} = \frac{Q_3 - Q_1}{2}$$

One way to measure how our numbers in Figure 10.3 vary is to consider how far away they are from their 'centre' (i.e. the arithmetic mean of 6). If we merely subtract the mean from each value to get the true deviation (some up and some down), we get numbers that because of their signs add up to zero and hence have a mean of zero. This must be so if the mean is their balance point. For our set of numbers (see Table 10.5), for example,

$$\sum f(x - \bar{x}) = -9 + 9 = 0$$

Table 10.5   Set 2: Calculation of mean average deviation ($\bar{x} = 6$)

x	f	Deviation $x - \bar{x}$	$f(x - \bar{x})$	$f\lvert(x - \bar{x})\rvert$
3	2	−3	−6	6
4	1	−2	−2	2
5	1	−1	−1	1
6	1	0	0	0
7	3	1	3	3
8	1	2	2	2
9	0	3	0	0
10	1	4	4	4
	10		0	18

$$\text{MAD} = \frac{\sum f\lvert(x - \bar{x})\rvert}{\sum f}$$

$$= \tfrac{18}{10} = 1.8$$

If, however, we consider that it is the amount of the deviation that matters (i.e. its *modulus* or numerical value) rather than whether it is up or down (+ or −), we can construct the *mean average deviation* (MAD), calculated to be 1.8 in Table 10.5. The two vertical lines | | indicate modulus; thus, for example, $|3-6| = |-3| = 3$. The MAD is of limited use as it is difficult to manipulate.

The deviation $(x - \bar{x})$ of each number from the mean is clearly a useful measure, but how else can we avoid the problem with signs?

Suppose that we square every deviation; this too removes negatives. Adding up these squared deviations (remembering that some have frequency greater than 1) and dividing by the total frequency gives us the mean of the squared deviations (called the *variance*):

$$\text{Variance} = \frac{\Sigma f(x-\bar{x})^2}{\Sigma f}$$

If we take the positive square root of this, to compensate for our squaring to remove negatives, we have the usual form for the most widely used measure of dispersion: the *standard deviation s*:

$$s = \sqrt{\frac{\Sigma f(x-\bar{x})^2}{\Sigma f}} = \sqrt{\text{Variance}}$$

This is calculated for our numerical example (set 2) in Table 10.6.

Table 10.6 *Set 2: calculation of standard deviation* ($\bar{x} = 6$)

x	f	$x - \bar{x}$	$(x - \bar{x})^2$	$f(x - \bar{x})^2$
3	2	−3	9	18
4	1	−2	4	4
5	1	−1	1	1
6	1	0	0	0
7	3	1	1	3
8	1	2	4	4
9	0	3	9	0
10	1	4	16	16
	10			46

$$s = \sqrt{\frac{\Sigma f(x-\bar{x})^2}{\Sigma f}}$$

$$= \sqrt{\tfrac{46}{10}} = \sqrt{4.6} = 2.14$$

Measures of Location and Dispersion   133

It is often not clear to students at this stage exactly what standard deviation measures, so it is worth repeating: *the standard deviation measures how spread out a set of numbers are around their mean.* Consideration of another set

$$\{4, 4, 5, 5, 6, 7, 7, 7, 7, 8\} \qquad \text{(Set 3)}$$

will show this. From Table 10.7 it can be seen that this set of numbers has the same mean as set 2, but the standard deviation is smaller, because the numbers are more closely bunched around their mean. If they were all equal to the mean, the standard deviation would be zero. (Incidentally, does this set also have the same mode and median as the previous set?)

Table 10.7   Set 3: Calculation of mean and standard deviation

x	f	fx	$x - \bar{x}$	$(x - \bar{x})^2$	$f(x - \bar{x})^2$
4	2	8	-2	4	8
5	2	10	-1	1	2
6	1	6	0	0	0
7	4	28	1	1	4
8	1	8	2	4	4
	10	60			18

$$\bar{x} = \frac{\Sigma fx}{\Sigma f}$$

$$= \frac{60}{10} = 6$$

$$s = \sqrt{\frac{\Sigma f(x - \bar{x})^2}{\Sigma fx}}$$

$$= \sqrt{\tfrac{18}{10}} = \sqrt{1.8} = 1.34$$

Notice that the definition of s is only interested in deviations (i.e. differences) and is not affected by absolute location. The numbers

$$\{83, 83, 84, 85, 86, 87, 87, 87, 88, 90\} \qquad \text{(Set 4)}$$

have a mean of 86 (they are all 80 more than those in set 2), but they have a standard deviation of 2.14, exactly the same as before, since

$$(x - \bar{x})^2 = (83 - 86)^2 = (-3)^2 = 9$$

just as

$$(x - \bar{x})^2 = (3 - 6)^2 = (-3)^2 = 9 \quad \text{etc.}$$

Note that an equivalent formula for s is

$$s = \sqrt{\left(\frac{\Sigma fx^2}{\Sigma f} - \bar{x}^2\right)} \quad \text{or} \quad \sqrt{\left(\frac{\Sigma fu^2}{\Sigma f} - \bar{u}^2\right)} \quad \text{(coded)}$$

This is more useful when $\bar{x}$ is not a nice round figure and the deviations $(x - \bar{x})$ would get complicated.

We shall finally work through the example of weekly incomes to find the standard deviation of grouped data (see Table 10.8). Once again, it is preferable to code the figures. The value of s_x is found by unscrambling the code, remembering that we don't need to add back the location figure, just magnify back to the old scale. This gives a standard deviation of £9.7 for the weekly income figures.

Table 10.8 *Weekly income: calculation of standard deviation ($\bar{u}$ = £3.15)*

$x(£)$	f	$u(£)$	u^2	fu^2
25	5	1	1	5
35	18	2	4	72
45	42	3	9	378
55	27	4	16	432
65	8	5	25	200
	100			1,087

$$s_u = \sqrt{\left(\frac{\Sigma fu^2}{\Sigma f} - \bar{u}^2\right)}$$

$$= \sqrt{\left(\frac{1{,}087}{100} - (3.15)^2\right)} = \sqrt{(10.87 - 9.92)}$$

$$= \sqrt{0.95} = 0.97 \quad \text{(i.e. £0.97)}$$

$$\therefore s_x = 0.97 \times 10 = 9.7 \quad \text{(i.e. £9.7)}$$

To calculate the quartile deviation or SIQ we need Q_1 and Q_3, where

$$Q_1 = \text{25th value} \quad \text{(of 100)}$$

$$Q_3 = \text{75th value}$$

Counting for Q_1, we reach cumulative frequency 23 at the beginning of class £40–49.99. We therefore need 2 more of the 42 values in that class, which has interval £10; hence,

$$Q_1 = 40 + (\tfrac{2}{42} \times 10) = 40.5 \quad \text{(i.e. £40.5)}$$

Similarly,
$$Q_3 = 50 + (\tfrac{10}{27} \times 10) = 53.7 \quad \text{(i.e. £53.7)}$$

$$\therefore \text{SIQ} = \frac{Q_3 - Q_1}{2}$$

$$= \frac{53.7 - 40.5}{2} = \frac{13.2}{2} = 6.6 \quad \text{(i.e. £6.6)}$$

Further divisions sometimes used are *deciles* and *percentiles*.

One final definition: the *coefficient of variation* V is the ratio between the standard deviation and the mean; that is,

$$V = \frac{s}{\bar{x}}$$

This is useful because both s and $\bar{x}$ have the original units, and V gives an indication of their relative size. We can compare different distributions using their coefficients of variation.

EXERCISES

(1) Given the figures in the table, calculate the values of $\sum_{i=1}^{8} x_i$, $\sum_{i=1}^{8} x_i y_i$, $\sum_{i=1}^{8} (x_i^2)$ and $\sum_{i=1}^{8} z_i$.

i	1	2	3	4	5	6	7	8
x_i	0	2	5	6	−7	4	−8	1
y_i	3	1	5	4	−9	6	2	8
z_i	3	3	3	3	3	3	3	3

(2) Use the table to calculate the average length of telephone calls to an exchange over a certain period.

Length of call (minutes)	Number of calls
0–2.99	13
3–5.99	7
6–8.99	10
9–11.99	9
12–14.99	5
Over 15	2

(3) Using the figures from Exercise 2 of Chapter 9, calculate the mean, median and mode order values.
Comment on the results.

(4) Using the figures from Exercise 3 of Chapter 9, calculate the mean, median and mode of the student examination marks.
Find also the quartile deviation of the marks.

(5) Two different-sized packets of detergent are filled by different machines. By calculating the mean and standard deviation of each of the following samples of packets filled by each machine, comment on their relative accuracy.

Contents of packet (in suitable units)	
Size A	Size B
22.7	27.8
23.1	26.2
21.8	26.1
22.2	25.1
22.6	25.9
22.4	26.7
	26.2
	26.8
	27.0

(6) The weekly sales of a particular type of confectionery are recorded by a shopkeeper over a year as shown in the array below. Form the data into a frequency distribution, using classes 30 to 39.99, 40 to 49.99, and so on.

36	50	69	60	74	63	83	66	57	48
68	77	63	58	64	39	67	53	72	75
81	97	74	62	44	63	73	41	68	62
56	65	41	53	68	57	69	79	32	71
75	89	34	64	84	61	75	92	86	51
45	78								

Calculate the mean weekly sales and the standard deviation of weekly sales.

Over what proportion of the year were sales within one standard deviation above and below the mean weekly sales?

What stock level could satisfy the weekly demand (sales) for 90 per cent of the year?

11 / Index Numbers

An *index number* is a statistical measure of change in a variable, or group of related variables, usually over time. The change can also be relative to a characteristic, such as income or profession.

In the last few years inflation at high levels has increasingly caused various forms of index numbers to enter the public consciousness. Index-linked saving schemes have recently been introduced by the government for restricted groups (mainly pensioners); and as this book is written, the idea of index-linked tax allowances is being openly proposed. The savings schemes, for example, provide an interest rate directly connected to the official rise in the cost of living rather than to a fixed rate.

The Retail Price Index is probably the best known. Another is the Index of Industrial Production, which uses production figures over the whole range of manufacturing industries. Other index applications may be to compare production levels of, say, steel in different regions of the country, or to compare the examination success of students in different years. A moment's thought will show that index numbers are of vital interest to trade union leaders (wages), farmers (production), economists (gold reserves and bank deposits) and accountants (depreciation estimates), as well as to the general public.

Let us consider a simple cost-of-living index such as is found in newspapers, where a cross-section of representative goods is taken at a particular (*base*) time. Often it is convenient for the whole 'shopping bag' to add up to a round figure like £5. The same goods, where possible, are then taken at a later (*current*) time; and by comparing the new total price with the old, an overall change can be measured. Suppose that the prices of a number of goods change as shown in Table 11.1. (The time periods could be weekly, yearly or anything else, depending on the context.) The cost-of-living index is thus 107/100. We usually express this as a percentage (i.e. 107).

It can be taken as an overall indicator of price rises because, although some goods have gone up and some have gone down (such fluctuations seem, like inflation, to be a feature of the world today), the selection of different quantities of each item fairly to represent its importance relative to the other items gives a true general picture. This relative importance

Table 11.1

Commodity	Price (pence) July 1977	July 1979
1 standard loaf of white bread	19	29
6 eggs	25	28
8 oz packet of margarine	15	17
1 lb of apples	33	25
1 lb of potatoes	8	8
	100	107

(called *weighting*) of each commodity is very important. We can arrive at it in a more complicated example for food by considering the expenditure (i.e. value) on each commodity by a typical family in a week. The unit for measuring each commodity (e.g. pints, pounds weight) is then irrelevant.

Over a long time a commodity may have to be included because it has become so important; for example, second-hand car prices and mortgages are much more important now than thirty years ago. There is a limit to the precision that is achievable, however, due to the expense and labour involved in the collection of data. Patterns of consumption also change; for example, coffee drinking and wine consumption may increase dramatically. Changes in quality may also affect comparability.

Let us now consider a longer example and some of the ways in which we can treat its figures.

Example 1

Petrol prices for the same month in 1976 and 1980, for the different grades, are given in Table 11.2.

We shall use the following notation for values, prices, quantities and weights respectively:

V, P, Q, W for the current period

v, p, q, w for the base period

We can construct a *price relative* for each grade of petrol:

e.g. $\quad \dfrac{P^{****}}{p^{****}} = \dfrac{134}{80} \times 100\% = 167.5\%$

If we also know the quantities sold for each grade of petrol in the two periods (see Table 11.3) we can also calculate a *quantity relative* for each grade. The changes could be due to restrictions on the lead content of petrol, safety factors for high performance cars (five-star) and the higher popularity of small cars (two-star):

Index Numbers

Table 11.2 *Price (pence/gallon)*

Year	**	***	****	*****
1976	77	78	80	82
1980	130	132	134	138

Table 11.3 *Quantity sold (in a suitable unit, e.g. millions of gallons)*

Year	**	***	****	*****
1976	2	2	5	1
1980	4	1.5	6	0.5

e.g. $\dfrac{Q^{*****}}{q} = \dfrac{0.5}{1} \times 100\% = 50\%$

The *value relative* is the product of the price relative and quantity relative, given by

$$\frac{V}{v} = \frac{P}{p} \cdot \frac{Q}{q}$$

e.g. $\dfrac{V^{****}}{v} = \dfrac{134}{80} \times \dfrac{6}{5}$

$= \dfrac{804}{400} \times 100\%$

$= 201\%$

We can construct a *value index* for all the grades, thus:

$$\frac{\text{Total value 1980}}{\text{Total value 1976}} = \frac{P_2 Q_2 + P_3 Q_3 + P_4 Q_4 + P_5 Q_5}{p_2 q_2 + p_3 q_3 + p_4 q_4 + p_5 q_5}$$

$$= \frac{(130)(4) + (132)(1\tfrac{1}{2}) + (134)(6) + (138)(\tfrac{1}{2})}{(77)(2) + (78)(2) + (80)(5) + (82)(1)}$$

$$= \frac{1591}{792}$$

$$= 2.009 \quad \text{(i.e. 200.9\%)}$$

Notice that for convenience we use numbers now instead of stars.

Alternatively, we can construct a *weighted aggregate* of prices, using as weights the quantities sold in the base period (i.e. 1976), thus:

$$\frac{P_2 q_2 + P_3 q_3 + P_4 q_4 + P_5 q_5}{p_2 q_2 + p_3 q_3 + p_4 q_4 + p_5 q_5}$$

To write such a summation of similar terms more neatly we use

$$\frac{\Sigma Pq}{\Sigma pq} \quad \text{or} \quad \frac{\Sigma Pw}{\Sigma pw} \qquad \text{(a \textbf{Laspeyres index})}$$

which simply means in each case 'add all the possible products of price times weight'; Σ (sigma) is the capital Greek S and denotes a sum. For the weighted aggregate index we get

$$\frac{\Sigma Pq}{\Sigma pq} = \frac{(130)(2) + (132)(2) + (134)(5) + (138)(1)}{(77)(2) + (78)(2) + (80)(5) + (82)(1)}$$

$$= \tfrac{1332}{792}$$

$$= 1.682 \qquad \text{(i.e. 168.2\%)}$$

This last index, using base year weights, is very often used. Since the weights in a *Laspeyres index* are base year weights, they only have to be established once. This is important, because repeated calculation of current weights is time consuming and expensive. However, base weights can quickly become irrelevant. The formula also tends to overstate a rise in prices, because it takes no account of the fall in output or demand that often goes with a large increase in price, or of the expansion of production that causes a fall in price.

If we choose to use current year weights, we shall have

$$\frac{\Sigma PQ}{\Sigma pQ} \quad \text{or} \quad \frac{\Sigma PW}{\Sigma pW} \qquad \text{(a \textbf{Paasche index})}$$

A *Paasche index* tends to understate the rise in prices, because it uses current weights. Although a Paasche index is generally preferable, a Laspeyres index is used if an index is required quickly or frequently.

Other variations on the same theme are as follows. Using a *typical year* method gives

$$\frac{\Sigma Pq_t}{\Sigma pq_t}$$

where q_t is the quantity for a typical year. If this typical-year quantity is taken as the arithmetic mean of the base and current quantities,

i.e. $q_t = \tfrac{1}{2}(q + Q)$

we get

$$\frac{\Sigma P(q + Q)}{\Sigma p(q + Q)} \qquad \text{(the \textbf{Marshall–Edgeworth index})}$$

Also, taking the *geometric mean* of Laspeyres and Paasche gives

$$\sqrt{\left(\frac{\Sigma Pq}{\Sigma pq} \cdot \frac{\Sigma PQ}{\Sigma pQ}\right)} \qquad \text{(Fisher's ideal index)}$$

Fisher's ideal index and the Marshall–Edgeworth index are used little in practice because of the calculations involved.

The introduction of our Σ notation now enables us to comment easily on some of the ideas so far mentioned.

Our value index was

$$\frac{\Sigma PQ}{\Sigma pq}$$

To construct a quantity index we do everything as before but interchange price and quantity; thus,

$$\text{Laspeyres index} = \frac{\Sigma Qp}{\Sigma qp}$$

$$\text{Paasche index} = \frac{\Sigma QP}{\Sigma qP}$$

A *simple aggregate* method uses

$$\frac{\Sigma P}{\Sigma p}$$

This is simple to use but takes no account of relative importance and is affected by the units used.

A *simple average of relatives* method (e.g. arithmetic mean) uses

$$\frac{\Sigma_n \frac{P}{p}}{N}$$

where the sum is taken over all the commodities, of which there are N. This still takes no account of relative importance, but it does not contain units.

A *weighted average of relatives* method weights each price relative by the total value of the commodity in some monetary unit:

(a) with base-year value weights

$$\frac{\Sigma\left(\frac{P}{p} \cdot pq\right)}{\Sigma pq} = \frac{\Sigma Pq}{\Sigma pq} \quad \text{(as before)}$$

(2) with current-year value weights

$$\frac{\Sigma\left(\frac{P}{p} \cdot PQ\right)}{\Sigma PQ}$$

(3) with typical-year value weights

$$\frac{\Sigma\left(\frac{P}{p} \cdot p_t q_t\right)}{\Sigma p_t q_t}$$

Example 2
Suppose that we have the prices for milk given in Table 11.4. Then we can express the price relatives for successive years as

$$p75/74 = \tfrac{8}{5} \times 100\% = 160\%$$
$$p76/75 = \tfrac{10}{8} \times 100\% = 125\%$$
$$p77/76 = \tfrac{12}{10} \times 100\% = 120\%$$

and for the whole period as

$$p77/74 = \tfrac{12}{5} \times 100\% = 240\%$$

These are known as *link or chain relatives*.

Table 11.4

Year	1974	1975	1976	1977
Price (pence/pint)	5	8	10	12

Example 3
Suppose that we want to change the base year, given the index numbers in Table 11.5. If we wish to make 1976 the base year (i.e. consider it as 100), we multiply every number through by 100/129. They are then all expressed as proportions of 129; for example,

Table 11.5

Year	1972	1973	1974	1975	1976	1977	1978	1979
Index	100	108	112	120	129	138	143	162

for 1977, 138 becomes $\frac{100}{129} \times 138 = 107.0$

for 1978, 143 becomes 110.9

for 1979, 162 becomes 125.6

Example 4

Real income is often not what it appears to be when measured relative to the past. Suppose that a person increases his salary by 62 per cent between 1975 and 1979 but the cost of living increases by 80 per cent in the same period. We need to deflate the series by considering income as a proportion of the cost of living; thus,

	1975	1979
Salary	100	162
Cost of living	100	180

Real income = $\frac{162}{180} \times 100\%$ = 90% of what it was

Example 5

We shall conclude the chapter by discussing in more detail the construction of a *retail price index*.

A new set of weights is established each year to cater for changes in expenditure patterns. Suppose that expenditure is divided into groups and the groups are weighted as shown in Table 11.6. These weights will probably have come from totals of expenditure in each category or group over the previous year and have been adjusted in proportion to one another to give a convenient total; for example, in 1976, alcohol expenditure at £5,980 million was 8.2% of total expenditure at £73,128 million.

Within each category, of course, there are many different items. Food, for example, needs to be broken down into meat, vegetables, dairy products, and so on, and each one of these needs to be further broken down. Different types of meat are at least all sold by the pound or kilogram, but to avoid the problem in dairy products, say, where units are not consistent, requires further care in producing the average dairy-products figure. For example, suppose that vegetables consists of potatoes, peas and

Table 11.6

Commodity group	Group subindex (relative to previous month)	Weighting
Food	102.5	350
Drink and tobacco	104.7	150
Clothing	101	105
Housing	100	90
Fuel and light	103.2	50
Household goods	101.3	65
Other goods and services	104	190
		1,000

mushrooms, priced and weighted as in Table 11.7. Weights are relative to quantities consumed; therefore,

$$\text{Average price of vegetables} = \frac{(9)(10) + (10)(4) + (35)(1)}{10 + 4 + 1}$$

$$= \frac{165}{15}$$

$$= 11\,\text{p}$$

Similarly, we can calculate the average price of other types of food, and we may obtain the prices shown in Table 11.8. Again, the weights will be based on the relative importance of each food type. We can now calculate an overall average price of food:

$$\text{Average price of food} = \frac{(90)(3) + (11)(5) + (48.4)(2) + \ldots}{3 + 5 + 2 + \ldots}$$

$$= \frac{421.8 + \ldots}{10 + \ldots}$$

$$= 41\,\text{p} \quad \text{(perhaps)}$$

Table 11.7

Vegetable type	Price (pence/lb)	Weighting
Potatoes	9	10
Peas	10	4
Mushrooms	35	1

Table 11.8

Food type	Average price (pence/unit)	Weighting
Meat	90	3
Vegetables	11	5
Dairy products	48.4	2
etc.	etc.	etc.

If this is compared to the food-category average price for the previous month of, say, 40p, we can obtain a subindex (i.e. relative) for food:

Food group subindex = $\frac{41}{40}$ × 100% = 102.5%

All the other categories or groups have their own subindices (i.e. relatives) compared to the previous month, as shown in Table 11.6.

Finally, the complete index can be calculated as a weighted average of price relatives by multiplying each group subindex by its own weight, adding up these products and dividing by the sum of the weights (i.e. 1,000). In this example we get

Retail price index = 102.7%

EXERCISES

(1) A polytechnic wants to plan the budget for the coming year with a forecast based on enrolments for the past year and estimated enrolments for the coming year. It is standard policy that 2 part-time degree students are equivalent to 1 full-time student, as are 3 part-time other-course students. With the figures given in the table, what should be the coming year's budget, if the past year's was £15 million?

Student type	Enrolments Past year (actual)	Coming year (estimated)
Full-time	3,650	4,400
Part-time (degree)	410	550
Part-time (other)	325	250

(2) Find the price relatives in December, with January as base, for each of the tabulated commodities used by an ice cream manufacturer to make his product in a certain year.

Month	Milk (pence/gallon)	Vanilla (£/jar)	Sugar (£/bag)	Edible oil (£/drum)
January	20	1.80	8.00	2.50
December	35	1.80	6.50	3.25

Use these relatives to find an (unweighted) average of relatives index number for the product.

(3) Calculate Laspeyres and Paasche indices for quantity of petrol, using the figures in Example 1 of the chapter.

(4) From the following figures obtain a new index of industrial production series, taking 1973 as base year.

Year	1968	1969	1970	1971	1972	1973	1974	1975	1976	1977	1978
Index	100	103	96	111	119	125	133	126	138	142	133

(5) Calculate the price relatives and quantity relatives for 1979, with 1969 as base, for the figures in the table.

Commodity	Price 1969	1979	Quantity 1969	1979
Milk	4.1	11.3	10,241	11,672
Butter	58.6	62.5	115	98
Cheese	37.4	58.8	81	89

Calculate the Laspeyres price index.

(6) How would you construct an index number for private expenditure on transport and vehicles?

(7) A factory uses three materials A, B and C in its manufacturing process. The prices are shown in the table. Using 1970 as a base year, calculate for 1980:

Material	Price 1970	(£/ton) 1980
A	8	9
B	120	112
C	70	83

(a) a simple aggregate price index.
(b) price relatives, and hence an average-of-relatives index.

Does either index suffer from any disadvantages?

If the number of tons of A, B and C used in 1970 were respectively 300, 50 and 100, calculate a weighted aggregate price index for 1980, using 1970 as the base year.

What purpose is served by using weights in the construction of an index?

(8) You are given the figures in the table. Explain the meaning of 'base 1977 = 100'.

Calculate an index of industrial production for 1979.

If the index for gas, electricity and water were to rise to 128 while the other figures remained the same, what would be the change in the index of industrial production?

Industry	Industrial production (base 1977 = 100) Weighting	Index 1979
Total manufacturing	748	104
Mining and quarrying	72	92
Construction	126	107
Gas electricity and water	54	113

12 Correlation and Regression

MEASURING THE DEGREE OF ASSOCIATION

We often meet a situation where data appear as pairs of figures relating to two variables, connected because each pair refers perhaps to a person or to the same year. For example, we may have the scores in an aptitude test and the subsequent sales performance for each of ten salesmen in a company, or the wheat yield corresponding to different applications of fertiliser in a series of trials.

If each pair of figures is plotted, we have what is called a *scatter diagram*, such as Figure 12.1 for the marks of 10 students in mathematics

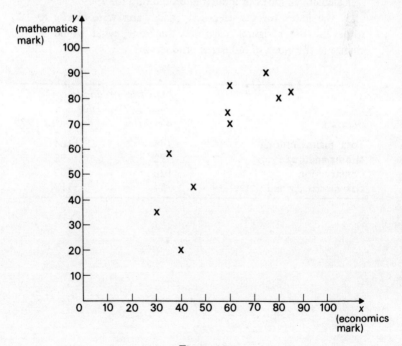

Figure 12.1

and economics examinations given in Table 12.1. Can we say that there is a link between achievement in the two subjects; and if so, how strong it is?

The degree of association between two variables is called the *coefficient of correlation r*. Strictly, this r is for linear associations. The various types of association are shown in Figure 12.2. Diagram (e) shows that we must be careful not always to assume a linear association.

Table 12.1

Student	A	B	C	D	E	F	G	H	I	J
Mathematics mark	45	20	35	58	85	90	80	82	75	70
Economics mark	45	40	30	35	60	75	80	85	58	60

We measure the coefficient of correlation r on a scale very much like that for probability, except that we need to distinguish between **negative correlations** (i.e. one variable increases as the other decreases) and **positive correlations** (i.e. both variables increase or decrease together). Thus,

$r = -1$ for perfect negative correlation

$r = 0$ for complete dissociation

$r = +1$ for perfect positive correlation

Most real-life examples lie somewhere between the extremes, and in Figure 12.2 we would get correlation coefficients (approximately in some cases) of:

(a) $r = 0$
(b) $r \simeq +0.9$
(c) $r \simeq -0.95$
(d) $r \simeq +0.7$
(e) $r \simeq 0$.

There are in fact two coefficients that we use. The first that we shall consider is *Spearman's rank correlation coefficient*. This is particularly useful if we need a quick calculation or if we do not have quantifiable variables; for example, we may be able to rank 8 apprentices in order of ability without having numbers for that ability. We usually use the symbol r' or r_s for this coefficient, and we need only to convert the figures or scores into order placings (called **ranks**). It is unimportant whether we rank upwards or downwards, as long as we do the same on each variable. I usually prefer to rank downwards, giving 1, 2, 3, and so on from the highest down.

With our examination-marks example we get the ranks shown in Table 12.2. Notice that equal ranking (E and J both have an Economics mark of

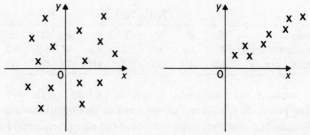
(a) Complete dissociation   (b) Strong positive linear association

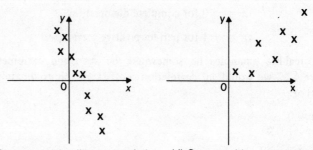
(c) Strong negative linear association   (d) Some positive association

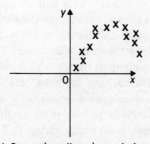
(e) Strong (non-linear) association

Figure 12.2

Table 12.2

Student	Mathematics ranking	Economics ranking	d	d^2
A	8	7	1	1
B	10	8	2	4
C	9	10	−1	1
D	7	9	−2	4
E	2	4.5	−2.5	6.25
F	1	3	−2	4
G	4	2	2	4
H	3	1	2	4
I	5	6	−1	1
J	6	4.5	1.5	2.25
				31.5

60) gives each 4.5, which is the arithmetic mean of 4th and 5th. Student I is then 6th. Symbol d represents the difference between the rankings of each pair. Sign is unimportant, as we square each d.

Now, r_s is defined as

$$r_s = 1 - \frac{6\Sigma d^2}{n(n^2 - 1)}$$

where n is the number of pairs. Clearly, if there is perfect positive correlation, each d will be 0 and r_s will be +1 as expected. In our example we have

$$r_s = 1 - \frac{6 \times 31.5}{10(10^2 - 1)}$$

$$= -\frac{189}{10 \times 99}$$

$$= -\frac{21}{110}$$

$$= -0.19$$

$$= 0.81$$

There is therefore a marked relationship between the variables.

The symbol r is usually reserved for the other, more complicated, coefficient. This is called the *product moment correlation coefficient*, defined as

$$r = \frac{n\Sigma xy - \Sigma x \Sigma y}{\sqrt{\{[n\Sigma x^2 - (\Sigma x)^2][n\Sigma y^2 - (\Sigma y)^2]\}}}$$

Each summation is calculated from the actual figures, calling one variable x and the other y, just as we did when calculating means and standard deviations in Chapter 10. We shall defer an example of the calculation of r until Example 1.

REGRESSION LINES

Once we know that two variables are strongly linearly related, it is natural to wonder if that relationship can be expressed as the equation of a line. We attempt to fit a 'line of best fit' (called a *regression line*) through the scatter diagram, but we need some criterion of 'best' to find its equation (see Figure 12.3). Clearly some measure of the closeness of the points to the line is needed.

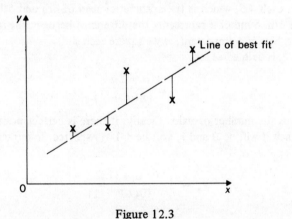

Figure 12.3

One obvious idea is to drop perpendiculars on to the line from each point and make the sum of the distances as small as possible. There are two objections:

(1) The variables are measured vertically and horizontally.
(2) Some distances will be positive (i.e. to points above the line) and others negative (i.e. to points below the line).

Consequently, we find the *line of regression of y on x* by adding up the *squares* of the *vertical* distances of each point from the line and making this sum as small as possible. We thus minimise the sum of squares.

The method is described in the section that follows, although you can

Derivation of Formulae for Regression Line Coefficients m and c

The regression line

$$y = mx + c$$

is shown in Figure 12.4, and the actual figures are represented by X and Y. A typical pair is (X, Y) at point P.

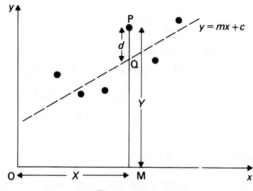

Figure 12.4

We attempt to minimise the sum of the squares of deviations (e.g. d) between the actual figure's y value at P and what the 'best' line gives as the y value, at Q; that is, we minimise Σd^2 for all d, where

$$d = PQ = PM - QM$$
$$= (y \text{ value at P}) - (y \text{ value at Q})$$

The y value at P is just Y, but at Q we find the y value from the regression line by substituting $x = X$ (which it does at M) into

$$y = mx + c$$

i.e. y value at $Q = mX + c$

$$\therefore d = Y - (mX + c)$$

$$\therefore \Sigma d^2 = \overset{\text{all } d}{\Sigma} d^2 = \overset{\text{all points}}{\Sigma} [Y - (mX + c)]^2$$

$$= \Sigma[Y^2 - 2Y(mX + c) + (mX + c)^2]$$

$$= \Sigma[Y^2 - 2mXY - 2cY + m^2X^2 + 2mcX + c^2]$$

We can now minimise by *partially* differentiating Σd^2 w.r.t. the unknowns m and c in turn. (Remember that X and Y are known for each plotted pair and are simply constants.) First, summing the r.h.s. terms separately, we get

$$D = \Sigma d^2 = \Sigma Y^2 - 2m\Sigma XY - 2c\Sigma Y + m^2\Sigma X^2 + 2mc\Sigma X + nc^2$$

since $\Sigma c^2 = nc^2$ for n pairs. Partial differentiation w.r.t c gives

$$\frac{\partial D}{\partial c} = 0 - 0 - 2 \cdot 1 \cdot \Sigma Y + 0 + 2m \cdot 1 \cdot \Sigma X + 2nc$$

$$= -2\Sigma Y + 2m\Sigma X + 2nc$$

For max./min. $\partial D/\partial c = 0$; that is

$$-2\Sigma Y + 2m\Sigma X + 2nc = 0$$
$$2nc = 2\Sigma Y - 2m\Sigma X$$
$$nc = \Sigma Y - m\Sigma X$$
$$c = \frac{\Sigma Y - m\Sigma X}{n}$$

Partial differentiation w.r.t. m gives

$$\frac{\partial D}{\partial m} = 0 - 2 \cdot 1 \cdot \Sigma XY - 0 + 2m \cdot \Sigma X^2 + 2 \cdot 1 \cdot c\Sigma X + 0$$

$$= -2\Sigma XY + 2m\Sigma X^2 + 2c\Sigma X$$

For max./min. $\partial D/\partial m = 0$; that is

$$-2\Sigma XY + 2m\Sigma X^2 + 2c\Sigma X = 0$$
$$2m\Sigma X^2 = 2\Sigma XY - 2c\Sigma X$$
$$m\Sigma X^2 = \Sigma XY - c\Sigma X$$

Substituting for c gives

$$m\Sigma X^2 = \Sigma XY - \Sigma X\left(\frac{\Sigma Y - m\Sigma X}{n}\right)$$

$$mn\Sigma X^2 = n\Sigma XY - \Sigma X\Sigma Y + m(\Sigma X)^2$$
$$mn\Sigma X^2 - m(\Sigma X)^2 = n\Sigma XY - \Sigma X\Sigma Y$$
$$m[n\Sigma X^2 - (\Sigma X)^2] = n\Sigma XY - \Sigma X\Sigma Y$$
$$m = \frac{n\Sigma XY - \Sigma X\Sigma Y}{n\Sigma X^2 - (\Sigma X)^2}$$

Incidentally,

$$\frac{\partial^2 D}{\partial c^2} = 2n \qquad \frac{\partial^2 D}{\partial m^2} = 2\Sigma X^2 \qquad (both\ positive)$$

$$\frac{\partial^2 D}{\partial c \partial m} = 2\Sigma X = \frac{\partial^2 D}{\partial m \partial c}$$

$$\therefore \frac{\partial^2 D}{\partial c^2} \cdot \frac{\partial^2 D}{\partial m^2} - \left(\frac{\partial^2 D}{\partial c \partial m}\right)^2 = 2n \cdot 2\Sigma X^2 - (2\Sigma X)^2$$

$$= 4n\Sigma X^2 - 4(\Sigma X)^2$$

$$= 4n\left(\Sigma X^2 - \frac{(\Sigma X)^2}{n}\right)$$

The r.h.s. is positive because it represents a squared deviation (compare your variance formula), so we do in fact have all the conditions for a minimum.

Use of the Regression Line

The end result of the above calculations is a formula that tells us the gradient m and the intercept c of this 'best' line of regression; and as we know from Chapter 1, they tell us everything about the line. We obtain the line of regression of y on x

$$y = mx + c$$

where

$$m = \frac{n\Sigma xy - \Sigma x \Sigma y}{n\Sigma x^2 - (\Sigma x)^2}$$

$$c = \frac{\Sigma y - m\Sigma x}{n}$$

Note that the formula for m is similar to that for r and also that c depends on knowing m.

The important thing about knowing the equation is that, given a value of x, we can predict or forecast a corresponding value of y.

In order to predict an x value from a given y value we need the *regression line of x on y*, which is slightly different. This is because it is derived by minimising the sum of squared *horizontal* deviations. It can be found from our formula by substituting x for y and y for x *everywhere*.

Two things in forecasting must be very carefully understood. First, the fact that there is a connection between variables does not necessarily mean that there is a *causal* relationship. It may be coincidence, or quite commonly they are both caused by some other variable. Secondly, we must also be careful when forecasting outside the limits of our information (called *extrapolating*), as the relationship may change or may only be approximately linear within our limits (see Figure 12.2e). It is usually safe to predict within our limits (i.e. to *interpolate*).

We shall now work through a full worked example.

Example 1

A study was made of the tendency of the height of a husband to be linearly related to that of his wife, and a random sample of 15 couples gave the results in Table 12.3.

Table 12.3

Couple	1	2	3	4	5	6	7	8	9	10	11	12	13	14	15
Wife's height (inches)	70	67	70	71	67	64	71	63	65	65	65	65	66	65	61
Husband's height (inches)	75	72	75	76	70	68	72	67	67	68	68	71	68	71	62

(a) Plot these results on a scatter diagram.
(b) Calculate the correlation coefficient and the least squares estimate of the regression line of husband's height on wife's height.
(c) Estimate the average height of husbands whose wives are 64 in. tall.
(d) How would you estimate the average height of wives whose husbands are 70 in. tall? Explain briefly why it would be incorrect to use the regression line that you have just calculated to do this.
(e) Discuss briefly practical factors that might lead the two variables to be causally related and hence those which would lead to a relationship that was *not* linear.

Solution:

(a) Let wife's height be x and husband's height be y (our prediction is this way round). The results are plotted in Figure 12.5.
(b) From the computations in Table 12.4, it can be seen that

$$\Sigma x = 995 \qquad \Sigma xy = 69,790 \qquad \Sigma x^2 = 66,127$$

$$\Sigma y = 1,050 \qquad \qquad \qquad \Sigma y^2 = 73,694$$

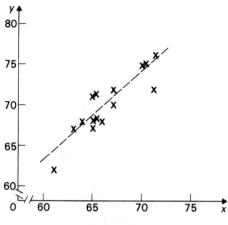

Figure 12.5

Table 12.4

x	y	xy	x^2	y^2
70	75	5,250	4,900	5,625
67	72	4,824	4,489	5,184
70	75	5,250	4,900	5,625
71	76	5,396	5,041	5,776
67	70	4,690	4,489	4,900
64	68	4,352	4,096	4,624
71	72	5,112	5,041	5,184
63	67	4,221	3,969	4,489
65	67	4,355	4,225	4,489
65	68	4,420	4,225	4,624
65	68	4,420	4,225	4,624
65	71	4,615	4,225	5,041
66	68	4,488	4,356	4,624
65	71	4,615	4,225	5,041
61	62	3,782	3,721	3,844
995	1,050	69,790	66,127	73,694

(A method for coding the variables, such as subtracting 60 in. from every figure, can be used, but with an electronic calculator my advice is to use the actual figures in a correlation and regression question. Any loss in time is made up by less risk of wrongly converting back two variables.) Hence, with $n = 15$,

$$r = \frac{15(69{,}790) - (995)(1{,}050)}{\sqrt{\{[15(66{,}127) - (995)^2][15(73{,}694) - (1{,}050)^2]\}}}$$

$$= 0.898 \quad \text{(a very strong association)}$$

For the regression line

$$y = mx + c$$

of y on x we get

$$m = \frac{15(69{,}790) - (995)(1{,}050)}{15(66{,}127) - (995)^2} = 1.12$$

$$c = \frac{1{,}050 - (1.12)(995)}{15} = -4.09$$

i.e. $\quad y = 1.12x - 4.09$

The negative intercept looks curious, but remember that the relation only holds for x above about 60 in.

(c) If $x = 64$ in.,

$$y = (1.12)(64) - 4.09 = 67.6 \text{ in.} \quad \text{(estimated)}$$

If we plotted the line of regression on to the scatter diagram, we could equally well read this estimate off the graph.

(d) An estimate of wives' heights from husbands' requires the regression line of x on y.

(e) The relationship is probably causal because men usually choose a wife who is shorter for social reasons, and she would perhaps be expected to be at least 2 in. shorter. It may be non-linear because tall men have a choice among women of any height (and may indeed prefer very small women), whereas short men are restricted in their choice.

If any women's liberation advocates are annoyed by the way this is expressed, the dependence is taken for convenience and from tradition. It could easily be expressed the other way round!

COEFFICIENT OF DETERMINATION

A measure of how satisfactory a regression line is can be obtained from the coefficient of determination D, which is the square of the coefficient of correlation r; that is,

$$D = r^2$$

It is often expressed as a percentage and shows the proportion of the variation in y accounted for, or explained by, the regression line; for example,

$$\text{if} \quad r = 0.8$$
$$\text{then} \quad D = 64\%$$

The rest must be due to other factors, chance, and so on.

One final point, brought out in Exercise 6, will serve as a hint. If the regression line is used to forecast the time series of a variable that fluctuates seasonally (or with a cycle), we can refine our prediction from the line by noting how much the actual figures in a particular season are up or down on average from the line and adjust a prediction from the line accordingly.

In general, we break a time series down into the linear model

$$A = T + S + R$$

The actual figure A is considered to be broken down into parts, such as the trend component T (provided by a regression line, for example), the seasonal component S and a residual component R (due to some unknown or random element). This breaking down (or decomposition) of a time series can also be done with more complicated models, such as the multiplicative model

$$A = TSR$$

EXERCISES

(1) The table gives wastage rates (as percentages) of students who entered universities and left without success in a certain period. Calculate both coefficients of correlation between the wastage rates in the Arts and Science faculties.

University	1	2	3	4	5	6	7	8	9	10	11	12
Arts	20	16	3	14	14	11	18	11	14	20	17	15
Science	21	23	3	12	17	9	19	19	14	22	12	22

(2) The state of repair of ten machines and the number of defective items that they produce in a week are shown in the table. Calculate Spearman's rank correlation coefficient, and comment on the result.

Machine	A	B	C	D	E	F	G	H	I	J
Weeks since service	4	3	8	6	14	7	20	12	9	12
Defective items	2	8	14	6	20	8	28	26	14	23

(3) A company collects data about its advertising expenditure, and the corresponding sales figures over a period of consecutive months, as shown in the table. Plot these figures on a scatter diagram.

Expenditure (£,00)	3	3.2	3.5	4	4.4	4.7	5.2	5.5
Sales (£,000)	7	8.3	9	10	10.5	10.8	11	11.1

Find the regression line of sales on advertising expenditure, and superimpose it on the scatter diagram.

What sales would you predict with an advertising expenditure of £600?

(4) Find the least-squares regression line of profits on capital employed for the companies in the electronics industry given in the table.

Capital employed (£ millions)	Profit (% of turnover)
3	9.4
4	8.3
5	11.2
5	13.5
6	12.2
7	14.0
8	12.1

Estimate the profits for a company with capital employed of £5 million.

Discuss why you might hesitate to estimate profits for a company that invested £11 million.

(5) An insurance company wishes to predict how its sales trainees will perform, from an aptitude test given at the beginning of their six-week training course. Records are kept of 10 salesmen's performance in their first year with the corresponding test score, and these are given in the table. Plot the scatter diagram for these data, and calculate the rank correlation coefficient.

Test score	48	56	58	64	66	72	78	82	84	90
Sales	44	54	44	52	58	60	66	56	66	64

Calculate the product moment correlation coefficient, and interpret the result.

Find the appropriate regression line to predict the first year sales of a trainee with a test score of 55.

(6) Quarterly sales figures of a retail store, recorded for 10 quarterly periods, are shown in the table. Use linear regression to find the 'line of best fit' to the data, and hence predict the sales in the 4th period of 1979.

Year	Quarter	Sales (£1,000 units)
1977	1	12
	2	14
	3	9
	4	16
1978	1	15
	2	16
	3	13
	4	21
1979	1	19
	2	22

13 Permutations and Combinations

PERMUTATIONS

Example 1
Consider 3 particular cards from a pack (e.g. the King, Queen and Jack of Hearts). In how many ways can these 3 cards be arranged in order?

Listing them, we have

K Q J	Q K J	J K Q
K J Q	Q J K	J Q K

There are thus 6 *permutations* (i.e. arrangements where order matters).

Another approach to the same answer that is often useful goes like this: the first card can be any 1 of 3, and each of these possibilities can be followed by (and is therefore multiplied by the number of) 2 cards remaining, whatever these are. Once the first 2 places are gone, the last remaining card is left to fill 1 place; hence,

$$\text{Number of ways} = 3 \times 2 \times 1 = 6 \quad \text{(as before)}$$

It is sometimes helpful to consider each position as a box.

This removal of one more possibility by the selection made at each step leads in general to strings of numbers multiplied together, with each number one less than the previous one, which we call **factorials**, symbolised by !; for example,

$$3! = 3 \times 2 \times 1 = 6$$
$$5! = 5 \times 4 \times 3 \times 2 \times 1 = 120$$

They are all therefore single numbers in their own right (i.e. a number of arrangements). Note that $1! = 1$ and that $0!$ is defined to have value 1.

The number of ways of arranging a general number of n objects is

$$\underbrace{n}_{\text{1st place}} \times \underbrace{(n-1)}_{\text{2nd place}} \times \underbrace{(n-2)}_{\text{3rd place}} \times \underbrace{\ldots}_{\text{etc.}} \times 3 \times 2 \times \underbrace{1}_{\text{last place}} = n!$$

If we arrange n objects that are not all unlike (e.g. p are alike and q others are alike), the p like objects can be arranged among their positions in $p!$ ways *within* a larger arrangement but will not make it different; similarly for the q like objects. Therefore, for n objects there are

$$\frac{n!}{p!q!} \text{ permutations}$$

when p are alike and q are alike; for example, for the letters of the word STATISTICS

$$\frac{10!}{3!3!2!} \text{ permutations}$$

Example 2

Suppose that we want to know the number of permutations of *any* 3 cards drawn from a pack.

By similar reasoning to that before, the first can be any one of 52, followed by the second, which is any one of the remaining 51; and each of these (52×51) ways can be followed by any one of the remaining 50. Hence, there are

$$52 \times 51 \times 50 \text{ permutations}$$

Now

$$52 \times 51 \times 50 = \frac{52 \times 51 \times 50 \times 49 \times 48 \times \ldots \times 2 \times 1}{49 \times 48 \times \ldots \times 2 \times 1}$$

$$= \frac{52!}{49!}$$

$$= \frac{52!}{(52-3)!}$$

In general, if r objects are chosen from n unlike (i.e. different) objects, there are

$$\frac{n!}{(n-r)!} \text{ permutations}$$

which is written ^nP_r.

Example 3

In how many ways can 10 books be arranged on a shelf, if 2 must remain apart?

The 10! ways of arranging 10 books without restriction are made up of those with the 2 books strictly kept together and those with them kept apart. Keeping 2 together really means 9 objects (i.e. 9! arrangements) but the 2 can be swopped over and still be together (doubling the number of ways); hence,

$$10! = 2(9!) + \text{(Number of ways apart)}$$

$$\begin{aligned}
\text{Number of ways apart} &= 10! - 2(9!) \\
&= 10(9!) - 2(9!) \quad \text{(since 10! is just 10(9!))} \\
&= (10-2)(9!) \\
&= 8(9!)
\end{aligned}$$

Example 4

How many ways are there of arranging 5 people around a round table?

Normally, we would have 5! ways (in a line); but because the table is round, position relative to the table is unimportant. The effect is as if 1 person were fixed with the other 4 arranged about this person. There are therefore 4! ways. (*Note*: Any arrangement in a ring follows the same ideas.)

COMBINATIONS

The difference between a permutation and a *combination* is that in a combination the order does not matter; for example, there is only 1 combination of King, Queen and Jack of Hearts.

When we chose 3 cards from a pack, we found 52!/49! permutations (see Example 2). Whatever the 3 cards were, we included their own 6 internal arrangements as being different (i.e. $3 \times 2 \times 1$ possible orderings of the 3 cards), so this number must be 6 *times larger* than the number of combinations. Therefore, for 3 cards drawn from a pack there are

$$\frac{52!}{49!3!} \text{ combinations}$$

In general, if r objects are chosen from n unlike objects, there are

$$\frac{n!}{(n-r)!r!} \text{ combinations}$$

which is written ^nC_r.

Since we are usually not concerned with the order of the chosen objects (e.g. in a hand dealt from a pack), we often refer to combinations as simply the number of ways of choosing r objects from n.

Can you see that we are really dealing with counting the number of subsets of a given set?

Example 5

In how many ways can a group of 4 be chosen from 7 people?
Solution:

$$\text{Number of combinations} = \frac{7!}{4!3!}$$

$$= \frac{7 \times 6 \times 5 \times 4 \times 3 \times 2 \times 1}{4 \times 3 \times 2 \times 1 \times 3 \times 2 \times 1}$$

$$= 35$$

Notice how a large cancelling can usually be made with the factorials. You may like to check the answer by giving the people letters and listing the foursomes.

EXERCISES

(1) Calculate: (a) $^5P_2$, (b) $^6C_4$, (c) $^6C_2$ and (d) $^3C_3$.

(2) In how many ways can the letters of the word ACCURACY be arranged?

(3) How many 3 letter arrangements can be made from the word MATHS? What is the probability of picking a real English word at random from all these?

(4) A shelf that holds 8 books is to be filled from a choice of 13 equal-size books. If these include *The Readable Maths and Statistics Book*, Volumes 1 and 2, which must stand at the ends, how many ways are there of filling the shelf?

(5) At a college meeting of 6 departments each department sends a group of 3 representatives. If each group sits together, how many different arrangements are there at a round table?

(6) In how many ways can a tennis foursome be chosen from 5 women and 3 men, so that there is always at least 1 man in the foursome?

(7) If the Manchester United first-team squad has 8 Scots, 6 Irishmen and

5 Englishmen, how many possible selections will have at least 1 player from each country?

(8) A hand of five cards is dealt from a pack without replacement. How many possible hands are there?

What is the probability of a full house, defined as 2 of a kind with three of a kind?

(9) Explain carefully why the term 'combination lock' is a misnomer.

Can you think of an example of a misnamed 'permutation'?

(10) A safe lock has 4 dials each with 10 digits, actually set one above the other but shown in a line in Figure 13.1 for clarity. How many different sequences are possible?

How many sequences have no digit occurring more than once?

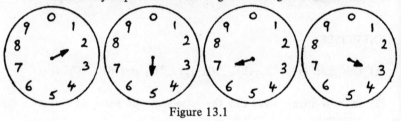

Figure 13.1

If we arrange n objects that are not all unlike (e.g. p are alike and q others are alike), the p like objects can be arranged among their positions in $p!$ ways *within* a larger arrangement but will not make it different; similarly for the q like objects. Therefore, for n objects there are

$$\frac{n!}{p!q!} \text{ permutations}$$

when p are alike and q are alike; for example, for the letters of the word STATISTICS

$$\frac{10!}{3!3!2!} \text{ permutations}$$

Example 2
Suppose that we want to know the number of permutations of *any* 3 cards drawn from a pack.

By similar reasoning to that before, the first can be any one of 52, followed by the second, which is any one of the remaining 51; and each of these (52 × 51) ways can be followed by any one of the remaining 50. Hence, there are

$$52 \times 51 \times 50 \text{ permutations}$$

Now
$$52 \times 51 \times 50 = \frac{52 \times 51 \times 50 \times 49 \times 48 \times \ldots \times 2 \times 1}{49 \times 48 \times \ldots \times 2 \times 1}$$
$$= \frac{52!}{49!}$$
$$= \frac{52!}{(52-3)!}$$

In general, if r objects are chosen from n unlike (i.e. different) objects, there are

$$\frac{n!}{(n-r)!} \text{ permutations}$$

which is written ^nP_r.

Example 3
In how many ways can 10 books be arranged on a shelf, if 2 must remain apart?

The 10! ways of arranging 10 books without restriction are made up of those with the 2 books strictly kept together and those with them kept apart. Keeping 2 together really means 9 objects (i.e. 9! arrangements) but the 2 can be swopped over and still be together (doubling the number of ways); hence,

$$10! = 2(9!) + (\text{Number of ways apart})$$

$$\begin{aligned}\text{Number of ways apart} &= 10! - 2(9!) \\ &= 10(9!) - 2(9!) \quad (\text{since } 10! \text{ is just } 10(9!)) \\ &= (10-2)(9!) \\ &= 8(9!)\end{aligned}$$

Example 4

How many ways are there of arranging 5 people around a round table?

Normally, we would have 5! ways (in a line); but because the table is round, position relative to the table is unimportant. The effect is as if 1 person were fixed with the other 4 arranged about this person. There are therefore 4! ways. (*Note*: Any arrangement in a ring follows the same ideas.)

COMBINATIONS

The difference between a permutation and a *combination* is that in a combination the order does not matter; for example, there is only 1 combination of King, Queen and Jack of Hearts.

When we chose 3 cards from a pack, we found 52!/49! permutations (see Example 2). Whatever the 3 cards were, we included their own 6 internal arrangements as being different (i.e. $3 \times 2 \times 1$ possible orderings of the 3 cards), so this number must be *6 times larger* than the number of combinations. Therefore, for 3 cards drawn from a pack there are

$$\frac{52!}{49!3!} \text{ combinations}$$

In general, if r objects are chosen from n unlike objects, there are

$$\frac{n!}{(n-r)!r!} \text{ combinations}$$

which is written ^nC_r.

Since we are usually not concerned with the order of the chosen objects (e.g. in a hand dealt from a pack), we often refer to combinations as simply the number of ways of choosing r objects from n.

14 The Binomial and Poisson Distributions

THE BINOMIAL DISTRIBUTION

Back in Chapter 7, on probability, we tentatively used a tree diagram to consider the outcomes when a coin is tossed twice (see Figure 7.1). Let us now tie together some of our ideas on probability, permutations and combinations, with an example of a family of 3 children.

Using the tree diagram, where each column represents one child (B = boy and G = girl), we get Figure 14.1. We thus have 8 permutations, such as BBG, which means a family of boy, boy, girl in that order of age.

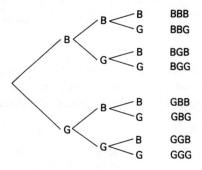

Figure 14.1

If we have no preference about age order and just want 2 boys and 1 girl, this combination is made up of 3 permutations:

$$BBG \quad BGB \quad GBB$$

∴ Probability of 2 boys and 1 girl = $\frac{3}{8}$

$$= 3(\tfrac{1}{2})(\tfrac{1}{2})(\tfrac{1}{2})$$

which is more easily written

$$p(2B, 1G) = \tfrac{3}{8}$$

(*Note:* We are assuming at the moment that boys and girls are equally likely to be born. We have a *binomial distribution*, because there are 2 possible outcomes on each *trial* (e.g. birth of a child, toss of a coin).

If we draw the histogram of the possible number of say, girls, in the family, we get Figure 14.2. Notice that 0 girls clearly implies 3 boys, and so on. Note also that the y axis, labelled as frequency, can just as easily be labelled probability, if every frequency is divided by the total number of outcomes (which is our definition of probability).

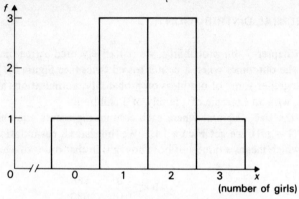

Figure 14.2

If we extend the tree diagram another stage by considering a fourth child in the family, we get 16 routes through the tree, corresponding to the 16 permutations shown in the histogram of different combinations (see Figure 14.3).

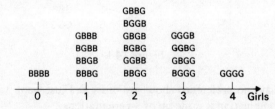

Figure 14.3

Notice that there are 6 ways of getting 2 girls and 2 boys. This corresponds exactly to our previous idea of choosing 2 objects from 4. There are $^4C_2 = 6$ ways of having 2 boys in a family of 4 children. What we are really doing is selecting from a product of brackets

$$(B + G)(B + G)(B + G)(B + G)$$

each of which represents a child. The linked arrows show one way of choosing (i.e. GBBG). If we were to multiply out $(B + G)^4$, we would get

$$(B + G)^4 = B^4 + 4B^3G + 6B^2G^2 + 4BG^3 + G^4$$
$$= {}^4C_0 B^4 + {}^4C_1 B^3 G + {}^4C_2 B^2 G^2 + {}^4C_3 BG^3 + {}^4C_4 G^4$$

which is a binomial expansion; $6B^2G^2$ corresponds to 6 combinations of 2 boys and 2 girls, for example. Hence,

$$p(2B, 2G) = 6(\tfrac{1}{2})^2(\tfrac{1}{2})^2 = \tfrac{6}{16}$$

Increasing the power to which we raise the bracket merely extends the process, and the coefficients of the various terms form a pattern known as **Pascal's triangle** (see Figure 14.4), which can be used to find a required coefficient. Each number is a number of combinations; for example, the 15 ringed is $^6C_2$ (i.e. the number of ways of combining 2 girls and 4 boys in a family of 6 children). Each number can also be obtained by adding the two directly above it, as shown with the arrows. For example, check that the next line is

$$1 \quad 7 \quad 21 \quad 35 \quad 35 \quad 21 \quad 7 \quad 1$$

Notice, if you have not already done so, the symmetry of the coefficients.

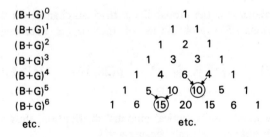

Figure 14.4 Pascal's triangle

The general statement of the binomial situation is: if we call the two outcomes success (S) and failure (F) and there are n independent trials, then the probability of r successes (and therefore of $n-r$ failures) in n trials is

$$^nC_r p^r (1-p)^{n-r}$$

where p is the probability of an individual success and $q = 1 - p$ is the probability of an individual failure. Note that n and p are called the *parameters* of the distribution. They specify it completely.

It is important for you to realise that the binomial distribution is the distribution of the probabilities of the various possible outcomes combining successes and failures in the n trials.

Our histograms in Figures 14.2 and 14.3 show that the shape of the distribution depends on n and on p. To show the dependence on p more clearly, let us consider different probabilities from those of

$$p(B) = p(G) = \tfrac{1}{2}$$

which gave rise to the two symmetrical distributions shown.

Suppose that 3 trials of an event are carried out, with

we have
$$p(\text{individual success}) = \tfrac{2}{5}$$
$$n = 3 \quad p = \tfrac{2}{5} \quad q = 1 - p = \tfrac{3}{5}$$

We can use the definition above; or alternatively, we can use the fourth line of Pascal's triangle and square this up with the corresponding combinations of successes and failures, as before, to get the expansion of $(S + F)^3$; that is,

$$\begin{array}{cccc} 1 & 3 & 3 & 1 \\ S^3 & S^2F & SF^2 & F^3 \end{array}$$

For a particular n it is the probability p that can change, *not* the number of combinations. The probabilities of the various outcomes are thus respectively

$$p(3S, 0F) = 1(\tfrac{2}{5})^3 = \tfrac{8}{125} \qquad p(2S, 1F) = 3(\tfrac{2}{5})^2(\tfrac{3}{5})^1 = \tfrac{36}{125}$$
$$p(1S, 2F) = 3(\tfrac{2}{5})^1(\tfrac{3}{5})^2 = \tfrac{54}{125} \qquad p(0S, 3F) = 1(\tfrac{3}{5})^3 = \tfrac{27}{125}$$

We therefore have a more typical binomial distribution, shown in Figure 14.5, which is not symmetrical, because $p \neq \tfrac{1}{2}$.

Example 1

A salesman of double glazing can conclude a sale on a call following an inquiry with a probability of $\tfrac{3}{10}$. What is the probability that 4 such calls in an evening conclude at least 2 sales?

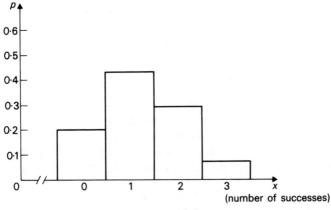

Figure 14.5

We know that

$$p(\text{individual success}) = \tfrac{3}{10}$$
$$\therefore p(\text{individual failure}) = \tfrac{7}{10}$$
$$n = 4$$

From the definition we get

$$p(2 \text{ sales in 4 calls}) = {}^4C_2(\tfrac{3}{10})^2(\tfrac{7}{10})^2$$
$$p(3 \text{ sales in 4 calls}) = {}^4C_3(\tfrac{3}{10})^3(\tfrac{7}{10})^1$$
$$p(4 \text{ sales in 4 calls}) = {}^4C_4(\tfrac{3}{10})^4(\tfrac{7}{10})^0$$

Since all these three satisfy the criterion, we add them to get

$$p(\text{at least 2 sales in 4 calls}) = 6\left(\frac{9}{100}\right)\left(\frac{49}{100}\right) + 4\left(\frac{27}{1{,}000}\right)\left(\frac{7}{10}\right)$$
$$+ 1\left(\frac{81}{10{,}000}\right) \cdot 1$$
$$= \frac{2{,}646 + 756 + 81}{10{,}000}$$
$$= \frac{3{,}483}{10{,}000}$$
$$= 0.3483$$

If the figures are complicated, we sometimes do not work through to an exact answer.

We could have got the same answer by our previous trick of the complement; that is

$$p(\text{at least 2 sales in 4 calls}) = 1 - [p(0 \text{ sales}) + p(1 \text{ sale})]$$

Suppose that the salesman makes 10 calls with the same probability $\frac{3}{10}$ of a sale on each call. What is his expected number of sales?

Intuition should suggest the answer 3, but what are we really finding? Imagine 10 calls together being made on a large number of occasions. The *expected value* is in fact the mean number of sales. In general,

$$\text{Mean} = np$$

where n is the number of trials and p the probability of success in each trial. Hence,

$$\text{Mean} = 10(\tfrac{3}{10}) = 3 \text{ sales}$$

Of course, he would not always get the mean value of sales — sometimes less and sometimes more — and the spread around the mean value is expressed in terms of the standard deviation, as before.

We state now, without proof, that for a binomial distribution

$$\text{Standard deviation} = \sqrt{[np(1-p)]}$$

where n and p are the parameters. We shall use this result later in the book.

It remains to discuss the case when n (i.e. total trials) and r (i.e. the number of successes) are so large that to calculate an answer exactly, as we have done in Example 1, is tedious. Fortunately, tables exist giving the probability of *r or more* successes in n trials for various convenient values of n and p. We obviously could not expect tables for all values. In Murdoch and Barnes we need Table 1 (page 4).*

The tables gives cumulative probabilities, because even with the individual probabilities worked out for us it would be boring to add up many of them when we require the probability of some number or more. Individual probabilities can easily be obtained by subtracting two successive cumulative probabilities; that is,

$$p(r \text{ successes}) = p(r \text{ or more}) - p(r+1 \text{ or more})$$

Example 2

If a production unit is made up from 20 identical components and each component has a probability of 0.25 of being defective what is the expected

*See Neave, page 5.

number of defective components in a unit? Also, what is the probability that in a unit:

(a) less than 3 components are defective?
(b) more than 3 components are defective?
(c) exactly 3 components are defective?

The expected number of defective components is, on average,

$$np = 20(\tfrac{1}{4}) = 5$$

Note: We are calling a defective a 'success' because 'success' is just a convenient label for whatever we are seeking.

First, we locate the section of Table 1 where $n = 20$ and $p = 0.25$. There are two sections for each n to allow different values of p: up to 0.1, and from 0.1 to 0.5. From the body of the table where $n = 20$ we look across to $p = 0.25$ and down to $r = 3$. (Remember that this really means $p(3 \text{ or more})$.)

(a) $\qquad p(3 \text{ or more}) = 0.9087 \qquad$ from the table

$\therefore p(\text{less than } 3) = 1 - p(3 \text{ or more})$

$\qquad\qquad\qquad\qquad = 1 - 0.9087$

$\qquad\qquad\qquad\qquad = 0.0913$

(b) $\qquad p(\text{more than } 3) = p(4 \text{ or more})$

$\qquad\qquad p(4 \text{ or more}) = 0.7748 \qquad$ from the table

$\therefore p(\text{more than } 3) = 0.7748$

(c) $\qquad p(\text{exactly } 3) = p(3 \text{ or more}) - p(4 \text{ or more})$

$\qquad\qquad\qquad\qquad = 0.9087 - 0.7748$

$\qquad\qquad\qquad\qquad = 0.1339$

If part (c) is difficult to grasp, think of the cumulative probabilities on an axis, as shown in Figure 14.6.

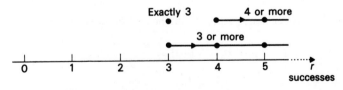

Figure 14.6

THE POISSON DISTRIBUTION

We define the *Poisson distribution* by saying that the probability of r successes in n trials is given by

$$p(r) = \frac{e^{-m}m^r}{r!}$$

where m is the average number of successes obtained in n trials.

Several things need explaining in this expression to avoid the onset of acute mathematical phobia for the student:

(1) r is playing its usual role of the particular number of successes asked for by ourselves or an exercise question. It will be given.
(2) $r!$ is the factorial of r, as before.
(3) $m = np$ as for the binomial distribution, where, as usual n is the number of trials and p is the probability of a success in one trial.

m may be given in the question. If not, we can calculate it from given data, as we calculated means before and shall do again in Example 3 below.

m^r is therefore just one number raised to the power of another number.

(4) The power in e^{-m} is $-m$ and will also be just a number, but what is this strange thing e?

Actually, e is just a constant approximately equal to 2.7183. The e stands for 'exponential'; and although it is a constant, it arises naturally in mathematics in ways that make it very important.

Consider as one example the function e^x, which after all is just like any other function of x, such as $(2x + 3)$, because it tells us what to do with any given value of x. The differential of e^x is equal to e^x. Thus, its rate of change is equal to itself — a unique property among functions (see Chapter 4).

For our purposes here with the Poisson distribution we simply look up in a table of e^x values or, more precisely, e^{-x} values, because we shall know the value of x (or m). Most sets of log tables include an e^{-x} table.

Let us consider some particular numbers of successes; for example,

$$\text{if} \quad r = 0, \quad p(0) = \frac{e^{-m}m^0}{0!} = e^{-m}$$

$$r = 1, \quad p(1) = \frac{e^{-m}m^1}{1!} = e^{-m}M$$

$$r = 2, \quad p(2) = \frac{e^{-m}m^2}{2!}$$

$$r = 3, \quad p(3) = \frac{e^{-m}m^3}{3!}$$

where $p(0) = p(0 \text{ successes})$, $p(1) = p(1 \text{ success})$, and so on. Notice that, once we know e^{-m}, it appears in every calculation.

The use of the Poisson distribution comes when we have situations where the probability of an individual success p becomes very small (i.e. close to zero) and the number of trials n becomes very large, but np (i.e. the mean m) stays fairly constant. Many 'rare event' situations in real life fit this closely (e.g. accidents in a factory, goals scored by football teams, faulty items off a production line). Some of the earliest work on the Poisson distribution was done in the nineteenth century on the number of deaths in Prussian cavalry regiments due to kicks by horses.

Example 3

Suppose that the goals scored by the 92 Football League teams on a particular Saturday are as given in Table 14.1.

Table 14.1

Number of goals, x	0	1	2	3	4	5	6
f (i.e. number of teams scoring that number of goals)	12	26	22	19	9	3	1

The histogram is shown in Figure 14.7, and the data are retabulated in Table 14.2. From the table we get

$$\text{Mean } \bar{x} = \frac{\Sigma xf}{\Sigma f}$$

$$= \frac{184}{92} = 2$$

$$\therefore \quad m = 2 \text{ goals}$$

using

$$p(r) = \frac{e^{-m}m^r}{r!}$$

with $m = 2$, we get

$$p(\text{a team scoring 0 goals}) = p(0) = \frac{e^{-2}2^0}{0!}$$

$$= 0.1353 \times 1 \quad \text{from tables of } e^{-\dot{x}}$$

$$= 0.1353$$

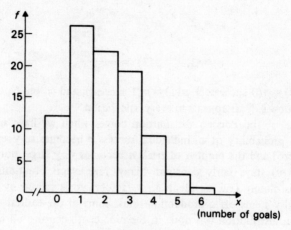

Figure 14.7

Table 14.2

x	f	xf
0	12	0
1	26	26
2	22	44
3	19	57
4	9	36
5	3	15
6	1	6
	92	184

That is, 92 teams would be expected to include (92 × 0.1353) teams scoring 0 goals,

i.e. 12.45 teams

We actually had 12. Also,

$$p(1) = \frac{e^{-2} 2^1}{1!}$$

$$= 0.1353 \times 2 = 0.2706$$

That is, there should be (92 × 0.2706) teams scoring 1 goal,

i.e. 24.9 teams

We actually had 26. Also,

$$p(2) = \frac{e^{-2}2^2}{2!}$$

$$= 0.1353 \times 2 = 0.2706$$

That is, 24.9 teams are expected to score 2 goals, against 22 in fact. Finally,

$$p(3) = \frac{e^{-2}2^3}{3!}$$

$$= 0.1353 \times \tfrac{4}{3} = 0.1804$$

That is, we expect (92 × 0.1804) teams to score 3 goals,

i.e.   16.6 teams

against 19 in fact.

You can see that the agreement is very good, which means that the Poisson distribution is a very good fit to the real life results.

An important use of the Poisson distribution is as an approximation to the binomial distribution, when p becomes small while n becomes large but np (i.e. m) stays constant.

The table for Poisson (Table 2, page 8, in Murdoch and Barnes)* is even simpler to use than the binomial table, because we only need one parameter, m. It is constructed in the same way with cumulative numbers of successes.

Example 4

The chance that a coalminer is killed in a mining accident in any one year is known from past records to be $\frac{1}{1800}$. Calculate the probability that a mine employing 540 miners has at least 1 fatal accident in a year.

We get

$$\text{Mean } np = 540 \times \tfrac{1}{1800} = 0.3$$

$$\therefore m = 0.3 \text{ deaths}$$

By calculation,

$$p(0 \text{ deaths}) = \frac{e^{-0.3}(0.3)^0}{0!}$$

$$= 0.7408 \times 1 \quad \text{from } e^{-x} \text{ tables}$$

$$= 0.7408$$

*See Neave, page 17.

$$\therefore p(\text{at least 1 death}) = 1 - 0.7408$$
$$= 0.2592$$

From Poisson tables

$$p(1 \text{ or more deaths}) = 0.2592 \quad \text{for } m = 0.3$$

If we were asked for the probability of, say, exactly 3 deaths, we would find

$$p(\text{exactly 3 deaths}) = p(3 \text{ or more}) - p(4 \text{ or more})$$
$$= 0.0036 - 0.0003 \quad \text{from Poisson tables}$$
$$= 0.0033$$

In an industrial situation the critical point is normally when the number of defectives per batch or sample is a certain number *or more*, so the cumulative table is particularly useful.

We have already stated many times that the mean of the Poisson distribution is m. We state now, without proof, that for the Poisson distribution

$$\text{Standard deviation} = \sqrt{m}$$

Compare this to the standard deviation of the binomial distribution which is $\sqrt{[np(1-p)]}$. You should be able to see that, as p becomes very small (i.e. closer to 0), $(1-p)$ gets closer to 1; that is,

$$p \to 0 \quad \text{and} \quad 1 - p \to 1$$
$$\therefore \sqrt{[np(1-p)]} \to \sqrt{np}$$
$$\text{i.e.} \quad \sqrt{[np(1-p)]} \to \sqrt{m}$$

Thus, the Poisson distribution approximates closer to the binomial distribution as p gets smaller (and n gets larger).

EXERCISES

(1) Use the complement method to complete as a check Example 1 about the salesman.

(2) If 5 trials are made of a situation, which can be a success (S) or failure (F), draw the tree diagram to show all the possibilities.

List all these possibilities; and by grouping them into combinations of success and failure draw the histogram of the distribution.

What is the most likely combination?

(3) If a biased coin is tossed 3 times, calculate the probability of at least 2 heads, given that on each toss

$$p(\text{head}) = \tfrac{2}{3}$$

(4) A marksman knows from past experience that he can hit a target 40% of the time. What is his probability of scoring at least 1 hit with 4 shots?

(5) In a certain country it is known that 15% of the population have blue eyes:

  (a) If 5 people are selected at random, what is the probability that:

   (i) 2 have blue eyes?
   (ii) none have blue eyes?

  (b) If a random sample of 100 people is taken, what is the probability that:

   (i) less than 10 have blue eyes?
   (ii) exactly 15 have blue eyes?

(6) If 35% of the cars in London are foreign makes, what is the probability that in a traffic jam of 20 cars there will be:

  (a) no foreign cars?
  (b) at least 1 foreign car?
  (c) exactly 5 foreign cars?

(7) Draw the histogram of all the various numbers of sales possibilities for Example 1 in the chapter.

(8) Using tables, check the probabilities for each of the numbers of goals as calculated in Example 3 of the chapter.

(9) Over 20 years 10 Prussian army regiments provided 200 observations of the number of deaths of cavalrymen due to kicks by their horse.

From the figures in the table, find the mean number of deaths, and calculate the probability of having two deaths per regiment per year. Check your answer using tables.

Number of deaths (per regiment per year)	0	1	2	3	4
Frequency	111	63	22	3	1

(10) A factory turns out screws of a standardised type in batches of 1,000. From experience it is known that on average 0.2% of each batch are defective. Find the probability that any batch includes more than 3 defective screws.

(11) A manufacturer knows from past knowledge that 0.5% of his product are faulty. They leave the factory in boxes of 200 items. If his guarantee replaces any box that contains more than 3 faulty items, at a cost of £10 per box to himself, what should he decide about a proposed inspection scheme that could eliminate all faulty items at a cost of 20p per box?

15 / The Normal Distribution

Throughout the book so far I have tried to introduce each mathematical concept after an example of a real situation to which it is applied or after at least a non-mathematical explanation of its meaning. In this chapter it is clearer to introduce the theoretical distribution first, so that you can learn to manipulate it freely before applying it to real problems.

The most important continuous distribution is the *normal distribution*, which is used as an approximation to many common real-life phenomena. It has the famous bell-shaped curve shown in Figure 15.1, where frequency is plotted against the value of a (continuous) variable in the usual way. It is a unimodal symmetrical curve with a complicated equation.

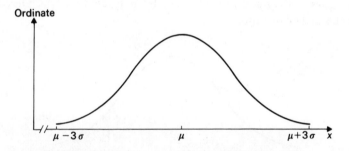

Figure 15.1 The normal distribution

We have previously had examples of frequency distributions in which we denoted the mean by $\bar{x}$ and the standard deviation by s. This was because they referred to samples of, for example, 40 students who were taken or selected in some way from the whole population of students. In order to distinguish between the cases of a sample and a population we use for the latter the notation of mean μ and standard deviation σ. The terminology is summarised in Table 15.1.

Notice in Figure 15.1 how the normal distribution has almost come down to touch the axis by the time it reaches 3 standard deviations either side of the mean; that is, there is negligible area under the curve beyond (i.e. outside)$(\mu + 3\sigma)$ and $(\mu - 3\sigma)$. There is nothing magical about this;

Table 15.1

	Sample	Population
Mean	$\bar{x}$	μ
Standard deviation	s	σ

it is merely something that is true about the dispersion of the normal distribution.

We shall see later that it is important to know areas under the curve. This has been done in a table (in Murdoch and Barnes, Table 3, page 13).* but we should first notice that different values for μ and σ will give squashed or stretched versions of the same shape, and we would need a special table for each case.

This problem is solved by *standardising*, rather like we did when coding a calculation. We create a new origin at the mean and remove the units of the variable by the transformation (code)

$$u = \frac{x - \mu}{\sigma}$$

This gives us the standard normal variable in Figure 15.2, which has mean 0 and standard deviation 1.

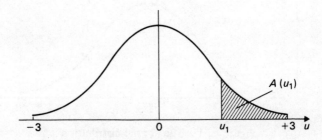

Figure 15.2

We shall work an example to show how this transformation is done a little later on, but let us first get used to handling the table.

Suppose that we want the area under the curve from $u = 1.16$ onwards (to $+$ infinity), as in Figure 15.3. We read the u value down the first column to the first decimal place (i.e. 1.1). (This is actually given in the Murdoch and Barnes table as the transformation for u, i.e. $(x - \mu)/\sigma$.) Then we read across to the second decimal place (i.e. 0.06) and find the area value corresponding to 1.16, which is 0.123.

This probably does not look much like an area to you, but all the areas are given as a *proportion* of the total area under the curve, which is

*See Neave, page 18.

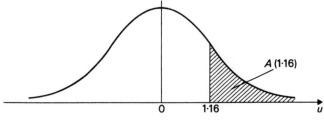

Figure 15.3

considered as 1. It is probably easier to think of this converted to percentages; thus, the total area is 100%, and

$$A(1.16) = 0.123 = 12.3\%$$

Therefore, 12.3% of the whole area under the curve lies to the right of 1.16, which is, remember, really a number of standard deviations because of our transformation to this standardised form of the normal distribution.

If we are interested in areas to the left of the mean, the symmetry of the curve allows us to look up the same figure for u, bearing in mind that it is negative. Therefore, the tail area below -1.16 is also 0.123 or 12.3 per cent. Check for yourself the area shown in Figure 15.4:

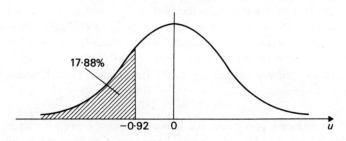

Figure 15.4

Area below $-0.92 = 17.88\%$

∴ Area above $-0.92 = 82.12\%$

You may come across tables that give the areas in a slightly different form. They may give the area between 0 and u, which for $u = 1.16$ would be 37.7%, or they may give the area up to u from below (from $-$ infinity), which is the complement of our own table areas and for $u = 1.16$ would be 87.7%. This complementary area is called $\Phi(\bar{u})$ in the table (see Figure 15.5).

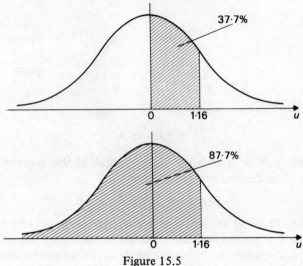

Figure 15.5

The only other thing that we shall ever need to read from our table is the reverse process when we know the area and need the mark corresponding to it.

Suppose that we want to know the number (of standard deviations) that has 25% of the area lying beyond it. We must convert 25% to the proportion 0.25, which we look for in the body of the table. The nearest we can get is between 0.2514 and 0.2483, which is between 0.67 and 0.68. Depending on the accuracy required, we split the difference; as it is virtually halfway, let us say 0.675 (see Figure 15.6). Hence, from tables

$$u_1 = 0.675 \text{ approx.}$$

As mentioned previously, the importance of the normal distribution is due to the number of natural phenomena that have approximately its shape. Particularly important among them are what are sometimes called *error distributions*. For example, items manufactured to a target (i.e. mean) dimension (e.g. the diameter of a piston) will not all have that target value but will instead have values above and below it. The majority will nonetheless often cluster closely around the target, and the number of them having larger errors of dimension will drop off dramatically as the dimension gets further away from the mean on each side. The result is our bell-shaped curve. Many human measurements like height and weight follow the same pattern, and we shall consider a typical case in Example 1.

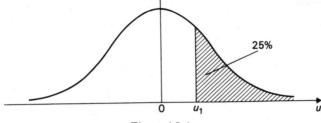

Figure 15.6

Example 1

Given that IQ (i.e. Intelligence Quotient) is approximately normally distributed, with mean 100 and standard deviation 12, how many of the population have IQ over 120? Also, what IQ is necessary to be in the top 1% of the population?

The figures here are fairly typical of a real situation, but it is very important to realise that we are strictly measuring the variable x(IQ) discretely; and even if we calculated the mean and standard deviation in the usual way for millions of people, we would still strictly have a histogram when we drew the frequencies against x. The normal distribution is a continuous curve and only approximates (although often extremely well) to our real distribution.

The two are shown together in Figure 15.7; and you should see how the area under the curve, beyond $x = 120$ for example, is just the number of people have IQ above 120, because each block in the histogram is a frequency or number of people on that mark. Adding all the blocks gives the whole area and *the frequency above 120* which can also be considered as a probability when expressed as a proportion of the total frequency or area, as our table is.

We have to convert 120 to the standardised normal. Strictly, 120 (as a discrete value) is the class mark of an interval between 119.5 and 120.5 (see histogram), and above 120 really means above 120.5. Sometimes, as in Example 2, we do not worry so much about this discrete measurement. Hence,

$$u = \frac{x - \mu}{\sigma}$$

$$= \frac{120.5 - 100}{12} = \frac{20.5}{12} = 1.71$$

From tables,

$$A(1.71) = 0.0436 \quad \text{(i.e. 4.36\%)}$$

That is, 4.36% of the population have IQ above 120.

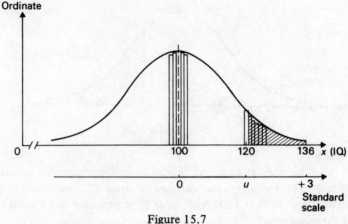

Figure 15.7

To find the IQ exceeded by 1% of the population we convert to the proportion 0.01 and find from the body of the table that the closest area is 0.0099, giving a u value of 2.33. Using the same transformation as above, but this time with x as the unknown instead of u, we get

$$2.33 = \frac{x - 100}{12}$$

$$27.96 = x - 100$$

$$x = 27.96 + 100$$

$$= 127.96 \quad \text{(rounded to 128)}$$

That is, an IQ of at least 128 is needed to be in the top 1% of the population.

We shall now work a slightly more complicated example.

Example 2

A company that manufactures batteries guarantees them a life of 24 months.

(a) If the average life has been found in tests to be 33 months, with a standard deviation of 4 months, how many will have to be replaced under guarantee if a normal distribution is assumed for the battery lifetimes?

(b) If annual sales are 10,000 batteries at a profit of £5 each, and each replacement costs the company £10, find the net profit.

(c) Would it be worth its while to extend the guarantee to 27 months, if the sales were to be increased by this extra offer to 12,000?

First, we standardise the variable in the usual way, using

$$u = \frac{x - \mu}{\sigma}$$

where

$$\mu = 33 \text{ and } \sigma = 4 \quad \text{(see Figure 15.8)}$$

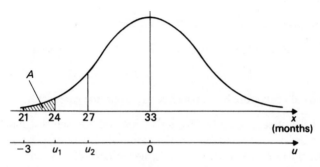

Figure 15.8

(a) Assuming that x is continuous, we get

$$u_1 = \frac{24 - 33}{4} = -\frac{9}{4} = -2.25$$

From tables,

$$A = 0.01222 \quad \text{(i.e. 1.222\%)}$$

Therefore, out of 10,000 batteries sold,

$$\frac{1.222}{100} \times 10,000 = 122.2 \text{ batteries}$$

will need replacing.

(b) Profit on sales of 10,000 = 10,000 × £5 = £50,000

  Replacement cost = 122.2 × £10 = £1,222

  ∴ Net profit = £50,000 − £1,222 = £48,778

(c) It is the balance of extra profit against extra replacement costs that matters. If the new guaranteed life is 27 months, we get

$$u_2 = \frac{27 - 33}{4} = \frac{-6}{4} = -1.5$$

From tables,

$$\text{Area below} -1.5 = 6.68\%$$

Therefore, if sales rise to 12,000 batteries,

$$\frac{6.68}{100} \times 12{,}000 = 801.6 \text{ batteries}$$

will need replacing. Hence,

New profit on sales of 12,000 = 12,000 × £5 = £60,000

New replacement cost = 801.6 × £10 = £8,016

∴ New net profit = £60,000 − £8,016 = £51,984

It would therefore be worthwhile to extend the guarantee.

We have already seen in Chapter 14 how the Poisson distribution can be used as an approximation to the binomial distribution. We shall now consider the normal approximation to the binomial distribution.

In Chapter 14 we noticed the shape of the binomial histogram for various small values of the number of trials n. Refresh your memory by looking at Pascal's triangle and drawing the histogram for $n = 6, p = \frac{1}{2}$.

As n increases, the shape of the histogram of the binomial distribution becomes more and more like that of the normal curve. We usually consider the approximation to be good enough if n is greater than 30 (and neither p nor q is too near 0). We have already stated that the mean and standard deviation of a binomial distribution are np and $\sqrt{[np(1-p)]}$ respectively, where n is the number of trials and p is the probability of success on one trial. We use this mean and standard deviation with a normal distribution for our approximation to the binomial, as follows in Example 3.

Example 3

If it is known that in a population 20% of the people vote for the Preservatives Party, what is the probability that in a random sample of 100 people there will be more than 25 Preservative voters?

We have $n = 100$ and $p = \frac{1}{5}$ (i.e. the individual probability of a Preservative voter). The 'expected' number of Preservative voters is given by the mean, where

$$\text{Mean} = np$$
$$= 100 \times \tfrac{1}{5} = 20$$

Standard deviation $= \sqrt{[np(1-p)]}$
$$= \sqrt{(100 \times \tfrac{1}{5} \times \tfrac{4}{5})} = 4$$

Assuming that the normal distribution approximation can be used, we get

$$u = \frac{x - \mu}{\sigma}$$

$$= \frac{25.5 - 20}{4} = \frac{5.5}{4} = 1.375$$

(Note the use of 25.5, because 25 is a discrete value). Interpolating from tables, we get

$$A(1.375) = 0.0846 \quad \text{(i.e. 8.46\%)}$$

That is, 8.46% of all samples contain more than 25 Preservative voters, *or* the probability that one sample contains more than 25 Preservative voters is 0.0846.

EXERCISES

Hint: A normal distribution diagram always helps.
Assume a standardised normal distribution in Exercises 1 to 4.

(1) Find the proportion of values that lie between:
  (a) 0 and + infinity.
  (b) 0 and 1.
  (c) 0 and 2.
  (d) 0 and 3.
  (e) −1 and 0.
  (f) −1.64 and 0.
  (g) −1.96 and −1.64.
  (h) −1.96 and 1.64.
  (i) −1.96 and 1.96.
  (j) −1 and +1.

(2) Find a value that has 95% of the area lying above it.

(3) Find a value that has 99% of the area lying below it.

(4) Find two values (symmetrically about 0) that include the middle 90% of the area.

Assume a population with a normal distribution having mean $\mu = 40$ units and standard deviation $\sigma = 2$ units in Exercises 5 to 8.

Note: A shorthand notation for such a normal distribution is

$$N(\mu, \sigma^2) = N(40, 2^2)$$

(5) What proportion of values lie between 41 and 41.5?

(6) What proportion of values are:
   (a) less than 41 units?
   (b) more than 41 units?

(7) Find two values symmetrically placed about μ such that they 'contain':
   (a) 90% of the population.
   (b) 95% of the population.

(8) If you chose a value at random from this population and came up with 45, would your confidence in the assumptions be disturbed?

(9) The mean in a certain normal population is 50. If 70% of the population exceed 47, what is the standard deviation?

(10) A bed manufacturer believes that the average height of men is 1.75 m, with a standard deviation of 5.8 cm. What length must he make the beds in order that no more than 0.5% of men are too tall for the beds manufactured? (Assume heights of men to be normally distributed.)

(11) A machine produces bolts to a mean length of 15 cm, with a standard deviation of 0.004 cm (assume a normal distribution). What percentage of bolts have lengths that lie outside the production tolerances of 15 ± 0.01 cm?

Due to wear the mean length-setting of the machine shifts to 15.002 cm. If the standard deviation remains unchanged, what percentage of bolts now fall outside the tolerances?

(12) Bags of sugar that contain a nominal 1 kg are filled by an automatic machine that is accurate to a standard deviation of 15 g. What should be the target setting for the machine to meet the legal requirement that only 2% of the bags may have contents of less than 1 kg?

A new and more accurate machine can reduce the variation of filling to a standard deviation of 10 g. What is the new target-setting required to meet the legal requirements still?

If weekly production is 100,000 bags and a bag containing G g has a production cost of

$$3 + \frac{G}{200} \text{ pence}$$

what is the weekly saving with the new machine, in sugar and in costs?

What would you take into account when deciding whether the new machine is a good investment?

(13) Boxes of Quantity Mix sweets contain 200 various-flavoured sweets after thorough mixing in production. If 1 in 8 of total sweet production is blackcurrant flavour, find the probabilities that a randomly selected bag contains:

(a) more than 30 blackcurrant sweets.
(b) between 15 and 30 blackcurrant sweets.

16 / The Probability Density and Cumulative Distribution Functions

REVISION: MEAN AND VARIANCE

At this stage we cannot do better than to revise with some simple examples one or two very important ideas developed earlier. They will be needed again and may in the meantime have become rusty in your memory. We can then expand these examples into further ideas that will smoothly transport us into the important concepts involved in statistical inference, the main concern of the statistics from here on, as opposed to our previous statistical work, which was largely descriptive.

Let us begin by refreshing our memories on the calculation of a mean and variance by calculating them for the scores that are possible on the throw first of one die and secondly (i.e. the total) of two dice. We shall also take the opportunity of introducing a slightly different notation that is sometimes used.

One Die

If x is the number thrown, the possible scores are obviously 1, 2, 3, 4, 5 and 6. Thus,

$$\text{Mean or expected score} = \frac{\Sigma x}{n}$$

$$= \frac{1+2+3+4+5+6}{6} = \frac{21}{6} = 3.5$$

and we can write this expected value as

$$E(x) = 3.5$$

The use of the expected value E as a synonym for the mean is simply because, in the long run, it is the number that you would expect to average or even out to. Remember how important gambling was historically to the development of probability theory.

To calculate the variance, remember that we first find the deviation of

The Probability Density and Cumulative Distribution Functions

each value from the mean, square to remove sign problems and then find the mean of these squared deviations. We had

$$\text{Variance} = \frac{\Sigma(x-\bar{x})^2}{n} \quad \text{or equivalently} \quad \frac{\Sigma x^2}{n} - (\bar{x})^2$$

and each had a version when frequencies were greater than 1 (see Chapter 10). Using our $E(x)$ notation and introducing $V(x)$ for the variance (compare our function notation previously), we get

$$V(x) = \frac{\Sigma[x-E(x)]^2}{n} \quad \text{or} \quad \frac{\Sigma x^2}{n} - [E(x)]^2$$

$$= E(x^2) - [E(x)]^2$$

This is best calculated in a table, as in Table 16.1.

Table 16.1: *calculation of variance (n = 6)*

x	$x - E(x)$	$[x - E(x)]^2$
1	−2.5	6.25
2	−1.5	2.25
3	−0.5	0.25
4	0.5	0.25
5	1.5	2.25
6	2.5	6.25
		17.50

$$V(x) = \frac{\Sigma[x - E(x)]^2}{n}$$

$$= \frac{17.5}{6} = 2.92 \quad \text{to 2 decimal places}$$

$$= 2\tfrac{11}{12} \quad \text{as a fraction}$$

Note: All frequencies are 1, so we need not bother further with f values.

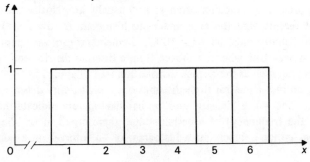

Figure 16.1

Remember this represents a measure of the spread of the x values about their mean, as shown in Figure 16.1 which is of course simply a histogram but shows that we have what is known for obvious reasons as a *rectangular distribution*.

As an aside, remember that we are really dealing with a hypothetical population of numbers generated by infinitely many throws of a die.

THE PROBABILITY DENSITY FUNCTION

This is a good chance to refresh your memory about the relationship between frequency and probability, namely, that

$$p = \frac{\text{Favourable outcomes}}{\text{Total outcomes}} = \frac{f}{n}$$

We can simply redraw the histogram with a (rescaled) y axis, as in Figure 16.2.

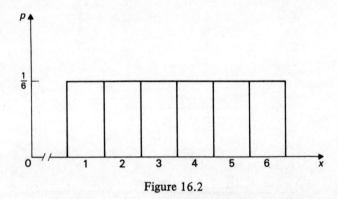

Figure 16.2

This may seem very simple, but the important things (which in practice are the ones that students often forget) usually are. Notice the very important feature that the area under our histogram is now 1, because the total probability must be 1 (or 100%). Remember also, very importantly, that this was true when we moved from a discrete distribution to a continuous one, such as the normal distribution (see Chapter 15).

The correct name for these distributions is *probability density function* or p.d.f. The words 'density' and 'probability' merely indicate the chance (really the frequency) of a certain value, represented in the diagram by an area, either a block on a histogram or an interval on a continuous curve, which sometimes is only approximating to a histogram (see Figure 16.3). The complete area under a p.d.f. must be 1.

The word 'function' just means that the probability (or frequency) is related to x in some way, just as before we had functions like

$$y = g(x)$$

The particular relationship will of course be implied in the question, such as the total score on a throw of two dice, which we can now deal with, belatedly but more rapidly, in a table.

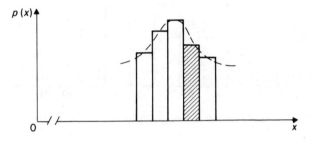

Figure 16.3

Table 16.2 *Two dice: Calculation of mean and variance*

x	f	p	fx	$x - E(x)$	$[x - E(x)]^2$	$f[x - E(x)]^2$
2	1	$\frac{1}{36}$	2	-5	25	25
3	2	$\frac{2}{36}$	6	-4	16	32
4	3	$\frac{3}{36}$	12	-3	9	27
5	4	$\frac{4}{36}$	20	-2	4	16
6	5	$\frac{5}{36}$	30	-1	1	5
7	6	$\frac{6}{36}$	42	0	0	0
8	5	$\frac{5}{36}$	40	1	1	5
9	4	$\frac{4}{36}$	36	2	4	16
10	3	$\frac{3}{36}$	30	3	9	27
11	2	$\frac{2}{36}$	22	4	16	32
12	1	$\frac{1}{36}$	12	5	25	25
	36		252			210

$$E(x) = \frac{\Sigma fx}{\Sigma f}$$

$$= \tfrac{252}{36} = 7$$

$$V(x) = \frac{\Sigma f[x - E(x)]^2}{\Sigma f}$$

$$= \tfrac{210}{36} = 5\tfrac{5}{6}$$

Two Dice

x is now the total score on two dice, and there are 36 possibilities. The mean and variance are calculated in Table 16.2. Notice that we could have used p instead of f to calculate $E(x)$ and $V(x)$, with the same result.

The probability density function in this case is not rectangular and looks like Figure 16.4.

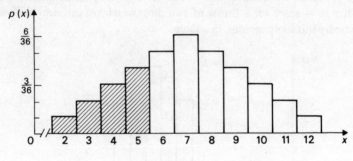

Figure 16.4

THE CUMULATIVE DISTRIBUTION FUNCTION

Suppose that we want not an individual probability but a cumulative probability, such as p (total of 5 or less), as shown shaded in Figure 16.4 to reinforce the idea of areas and probabilities. Whereas,

$$\text{if} \quad x = 5 \quad (\text{say})$$
$$\text{then} \quad p(x) = \tfrac{4}{36}$$

we now have, adding the four possibilities,

$$p(x \leqslant 5) = \tfrac{1}{36} + \tfrac{2}{36} + \tfrac{3}{36} + \tfrac{4}{36} = \tfrac{10}{36}$$

The proper name for the relationship between these cumulative probabilities and the corresponding x value is the distribution function $F(x)$, where

$$F(x) = p(x \leqslant x)$$

We have

$$F(5) = p(x \leqslant 5) = \tfrac{10}{36}$$

Similarly,

$$F(3) = p(x \leqslant 3)$$
$$= \tfrac{1}{36} + \tfrac{2}{36} = \tfrac{3}{36}$$

If we work out the others in the same way, we can draw this distribution function (really just a cumulative probability), as in Figure 16.5.

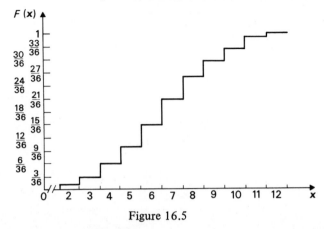

Figure 16.5

This is known as a **step function**, because each cumulative probability is jumping or stepping up to its next value.

The area situation in Figure 16.4 is of course completely equivalent in the continuous case, such as the normal distribution. If you look at Table 3 (page 13) of Murdoch and Barnes,* you should remember that the table probability gives the tail areas, which are the complement of the cumulative areas, expressed there as $\Phi(u)$. Some tables, as previously mentioned, actually give values of $\Phi(u)$ instead of $[1 - \Phi(u)]$ and are therefore a complete list of the distribution function. The areas are calculated in a continuous case using the integral calculus (see Chapter 6).

As one more example of a distribution function, consider the general rectangular distribution:

$$p = f(t) = \begin{cases} 0 & \text{if } t < a \\ \dfrac{1}{b-a} & \text{if } a \leqslant t \leqslant b \\ 0 & \text{if } b < t \end{cases}$$

It looks like Figure 16.6. Notice that the area underneath is unity. Its distribution function looks like Figure 16.7.

Let us finish this mainly revision chapter by considering an example of a continuous probability density function. Remember that a p.d.f. specifies the relationship between a variable and the probability of certain values of that variable and that the p.d.f. very largely summarises the population that the variable covers. A diagram of the p.d.f. gives a picture summary of the population. We shall not consider a p.d.f. as complicated as, say, the normal distribution, but we shall extract typical characteristics from the following example.

*See Neave, page 18.

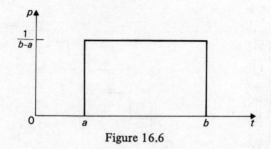

Figure 16.6

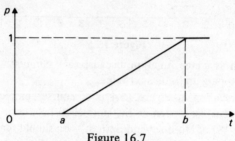

Figure 16.7

Example 1

The p.d.f. of a variable x, which has values between 0 and 10, rises linearly to a maximum at $x = 6$ and then decreases linearly to 0 at $x = 0$. Find:

(a) the maximum value of p.
(b) the proportion of the population that is:
 (i) less than 2.
 (ii) between 1 and 4.
 (iii) between 6 and 10.
(c) the mode.
(d) the median.

The p.d.f. is plotted in Figure 16.8.

(a) We let the maximum value of p (at the value $x = 6$) be h. Then, because the total area under a p.d.f. must be 1, the areas of the two triangles ① and ② must sum to 1; we get

$$1 = \tfrac{1}{2} \cdot 6h + \tfrac{1}{2} \cdot 4h$$
$$= 3h + 2h$$
$$= 5h$$
$$h = \tfrac{1}{5}$$

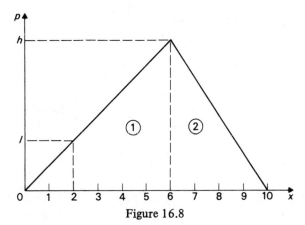

Figure 16.8

(b) (i) By similar triangles, the probability l at $x = 2$ is given by the proportion

$$\frac{l}{h} = \tfrac{2}{6}$$

$$\therefore l = \tfrac{2}{6} h$$

$$= \tfrac{2}{6} \cdot \tfrac{1}{5} = \tfrac{1}{15}$$

(Remember that a continuous variable only has meaningful probabilities in an interval or range of values of x. At a particular value the probability will be zero). Between $x = 0$ and $x = 2$ the area under the p.d.f. is contained in a triangle; hence,

$$\text{Area} = \tfrac{1}{2} \cdot 2l$$

$$= \tfrac{1}{2} \cdot 2 \cdot \tfrac{1}{15} = \tfrac{1}{15}$$

That is, $\tfrac{1}{15}$ of the population is less than 2.

(ii) Similarly, by similar triangles we find that the area under the p.d.f. from $x = 1$ to $x = 4$ is contained in a trapezium whose parallel sides are respectively $p = \tfrac{1}{30}$ and $p = \tfrac{2}{15}$; hence,

Area = $\tfrac{1}{2}$ (Sum of parallel sides)(Distance between them)

$$= \tfrac{1}{2} (\tfrac{1}{30} + \tfrac{2}{15})(4 - 1)$$

$$= \tfrac{1}{2} \cdot \tfrac{5}{30} \cdot 3$$

$$= \tfrac{1}{4} \quad \text{(i.e. 25\%)}$$

That is, 25% of the population lie between 1 and 4.

(iii) The area under the p.d.f. from $x = 6$ to $x = 10$ is simply that of triangle ②;

$$\text{Area} = \tfrac{1}{2} \cdot 4h$$
$$= \tfrac{1}{2} \cdot 4 \cdot \tfrac{1}{5} = \tfrac{2}{5} \quad \text{(i.e. 40\%)}$$

That is, 40% of the population are between 6 and 10.

(c) The mode is the maximum value of p, found in (a) (i.e. $\tfrac{1}{5}$).

(d) The median value of x has 50% of the frequencies (i.e. the area) above it and below it. Since we know that triangle ② has 40% of the area, the median m must split triangle ① as shown in Figure 16.9. By proportion,

$$\frac{m}{6} = \frac{H}{\tfrac{1}{5}}$$

$$H = \frac{m}{6} \cdot \tfrac{1}{5}$$

$$= \frac{m}{30}$$

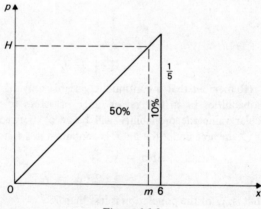

Figure 16.9

The areas of the triangles in Figure 16.9 give

$$\frac{\tfrac{1}{2} \cdot mH}{\tfrac{1}{2} \cdot 6 \cdot \tfrac{1}{5}} = \tfrac{50}{60}$$

$$\frac{mH}{\tfrac{6}{5}} = \tfrac{5}{6}$$

$$mH = \tfrac{6}{5} \cdot \tfrac{5}{6}$$

$$= 1$$

This is correct, because $\frac{1}{2}mH$ represents 50% of the total. Substituting for H gives

$$m \cdot \frac{m}{30} = 1$$

$$m^2 = 30$$

$$m = \sqrt{30}$$

$$= 5.48$$

EXERCISES

(1) x is the random variable showing the number of boys in families with 4 children. Construct a table of the probability distribution, and find the mean and variance of the distribution.

Draw a diagram of the distribution function. Assume that

$$p(\text{boy}) = \tfrac{1}{2}$$

(2) The (infinite) population of x consists of the numbers 0, 1, 2 and 3 in the proportions $1:4:2:1$ respectively. Find:

(a) $p(x)$ for $x = 0, 1, 2, 3$.
(b) the probability that an x chosen at random is greater than or equal to 2.

Draw the distribution function $F(x)$.

(3) The p.d.f. for a variable x, which varies from 0 to 40, rises linearly from 0 at $x = 0$ to h at $x = 40$. Draw the p.d.f., and find:

(a) the value of h.
(b) the proportion of xs greater than 20.
(c) the probability that a member chosen at random from the population will show a value of x between 20 and 30.

(4) A (discrete) variable has probability (density) function

$$p(x) = k(\tfrac{1}{3})^x$$

where $x = 0, 1, 2, \ldots$. Sketch the function, and:

(a) find k.
(b) calculate $p(2)$.
(c) find the probability that $x \leq 3$.

Hint: The sum of an infinite geometric progression $a, ar, ar^2, \ldots$, where $r < 1$, is

$$\frac{a}{1-r}$$

(5) Check the results in the chapter for the variance of the scores on one die and (the total) on two dice, using the alternative formula for variance, namely,

$$E(x^2) - [E(x)]^2$$

17 Samples and the Standard Error

We need to move away now from merely describing data and organising it. We want to use the numerical quantities that we have constructed in describing the location and dispersion of a sample (each numerical quantity is known as a statistic) in order to estimate numerical constants characterising the population (these are known as parameters). We can thus, it is hoped, make inferences from the statistics to the parameters (i.e. from the sample to the population).

Just as one example, if it is impossible to measure the mean diameter of every piston produced by a large car manufacturer, can we get a good idea of what it is by measuring the mean diameter of a sample of pistons?

The populations under consideration can be real or hypothetical, and small, large or infinite. It is of course essential to be able to assume that the population is represented in miniature by the sample during large scale sampling. The problems of sampling, the importance of randomness, practical methods, questionnaires, and so on are left until the later chapter on sampling (Chapter 24). For now we shall try to continue a smooth flow of ideas by considering a more simple and theoretical example.

MEAN AND VARIANCE OF SAMPLE MEANS (INFINITE POPULATION)

Example 1

A hat contains 4 pieces of paper with the numbers 2, 2, 3 and 5 written on them. A sample of 2 pieces of paper is chosen at random with replacement; that is, the first piece is replaced before the second is chosen. Find the mean and variance of the original population, and the mean and variance of the sampling distribution of the means of the samples of 2 numbers.

First, we assume that this is a practical version of an **infinite population** of equal amounts of the numbers 2, 2, 3 and 5.

Secondly, **with replacement** means that the number drawn first can possibly be drawn again in the second draw. **Without replacement** (which is usually the case in practical sampling) it obviously stays out and cannot reappear.

The original population can be set up theoretically using a table, from which population mean μ and variance σ^2 can be calculated (see Table 17.1).

Suppose that we take a sample of 2 numbers and get 2 and 5. Their mean is 3.5. Now, suppose that we take another sample and get 5 and 3, which have mean 4. Clearly, we could go on taking samples of size 2 and find the mean each time. Generally, this sample mean will be different each time; but although we could continue indefinitely and find what is called the *sampling distribution* of means of samples size 2 from the population, there are only a limited number of possibilities. These would arise in huge equal numbers or can be thought of more elegantly as forming a theoretical distribution, which we can either list one by one or express in a table. Each number 2, 2, 3 or 5 can appear first with any one of the same appearing second. Halving the total of each pair (i.e. sample of 2) gives the corresponding sample mean $\bar{x}$ (see Table 17.2).

Table 17.1 *Population: calculation of mean and variance*

x	f	fx	$x - \mu$	$(x - \mu)^2$	$f(x - \mu)^2$
2	2	4	-1	1	2
3	1	3	0	0	0
5	1	5	2	4	4
	4	12			6

$$\mu = \frac{\Sigma fx}{\Sigma f}$$

$$= \tfrac{12}{4} = 3$$

$$\sigma^2 = \frac{\Sigma f(x - \mu)^2}{\Sigma f}$$

$$= \tfrac{6}{4} = 1.5$$

Table 17.2 *Sample means $\bar{x}$*

	Second draw			
	2	2	3	5
First draw — 2	2	2	2.5	3.5
2	2	2	2.5	3.5
3	2.5	2.5	3	4
5	3.5	3.5	4	5

Can you see that without replacement we would get the same table less the diagonal boxes? (See Table 17.6.) It would lead to a completely different mean and variance.

Calculating with our usual table we get the mean $\mu_{\bar{x}}$ and variance $\sigma_{\bar{x}}^2$ of the sample means (see Table 17.3).

Table 17.3 *Sample means: calculation of mean and variance*

$\bar{x}$	f	$f\bar{x}$	$\bar{x} - \mu_{\bar{x}}$	$(\bar{x} - \mu_{\bar{x}})^2$	$f(\bar{x} - \mu_{\bar{x}})^2$
2	4	8	−1	1	4
2.5	4	10	−0.5	0.25	1
3	1	3	0	0	0
3.5	4	14	0.5	0.25	1
4	2	8	1	1	2
5	1	5	2	4	4
	16	48			12

$$\mu_{\bar{x}} = \frac{\Sigma f \bar{x}}{\Sigma f}$$

$$= \tfrac{48}{16} = 3$$

$$\sigma_{\bar{x}}^2 = \frac{\Sigma f(\bar{x} - \mu_{\bar{x}})^2}{\Sigma f}$$

$$= \tfrac{12}{16} = 0.75$$

MEAN OF SAMPLE VARIANCES (INFINITE POPULATION)

Now, a statistic used to estimate a parameter is known as an *estimator*; for example, in Example 1 our first sample mean of 3.5 was an estimate of the population mean, which happened to be 3. What properties should an estimator possess to be considered a good estimator? Well, we hope that these estimates of 3.5, 4, and so on as we take more samples will centre around the parameter 3, rather like bullet holes round a target bullseye. More precisely, we want the average (i.e. mean) value of the sample statistic to equal the population parameter that it is estimating. Statistics that have this property are called *unbiased estimators* of the parameter.

In Example 1 the mean of the sample means was equal to the population mean (i.e. both 3). It is also true in general that the sample mean is an unbiased estimator for the population mean. The same is true for the median.

We did not of course calculate the average (i.e. mean) value of the sample variances. Let us consider the sample of 5 and 3 as an illustration:

Sample 5 and 3

Sample mean $\bar{x} = 4$

Sample variance $\dfrac{\Sigma(x-\bar{x})^2}{n} = \dfrac{(5-4)^2 + (3-4)^2}{2}$

$= \dfrac{1+1}{2}$

$= 1$

You are asked in Exercise 1 to check the variance for every sample in Example 1, but the complete list is given in Table 17.4. The mean of these sample variances (i.e. $\tfrac{13}{16}$) is not equal to the population variance (see Table 17.1). Therefore,

$$\dfrac{\Sigma(x-\bar{x})^2}{n}$$

is a **biased estimator**.

Table 17.4 *Sample variances; calculation of mean*

Sample	Sample mean,	f	Sample variance,	f × Sample variance
2, 2	2	4	0	0
2, 3	2.5	4	0.25	1
3, 3	3	1	0	0
2, 5	3.5	4	2.25	10
3, 5	4	2	1	2
5, 5	5	1	0	0
		16		13

Mean of sample variances $= \dfrac{\Sigma(f \times \text{Sample variance})}{\Sigma f}$

$= \tfrac{13}{16}$

Actually, it gives a value that is too low, and it is customary to calculate the sample variance using

$$\dfrac{\Sigma(x-\bar{x})^2}{(n-1)}$$

instead. By dividing by $(n-1)$ instead of n we make each sample variance larger. Recalculating the sample variance of the illustration, we get:

Sample 5 and 3

Sample mean $\bar{x} = 4$

Sample variance $\dfrac{\Sigma(x-\bar{x})^2}{n-1} = \dfrac{(5-4)^2 + (3-4)^2}{1}$

$= 1 + 1$

$= 2$

Recalculating for every sample, we get the new mean of sample variances to be $\frac{24}{16}$ or 1.5 (see Table 17.5), which is equal to the population variance. Therefore, the statistic

$$\dfrac{\Sigma(x-\bar{x})^2}{n-1}$$

of sample variance is an *unbiased estimator* for the parameter of population variance.

Table 17.5 *Sample variances: calculation of mean*

Sample	Sample mean,	f	Sample variance,	f × Sample variance
2, 2	2	4	0	0
2, 3	2.5	4	0.5	2
3, 3	3	1	0	0
2, 5	3.5	4	4.5	18
3, 5	4	2	2	4
5, 5	5	1	0	0
		16		24

New mean of sample variances $= \dfrac{\Sigma(f \times \text{Sample variance})}{\Sigma f}$

$= \frac{24}{16} = 1.5$

Note that, as sample size n increases the difference that this makes in the denominator becomes less important.

VARIANCE OF SAMPLE MEANS AND STANDARD ERROR

We should now consider the value that we obtained for the variance of the sampling distribution of the sample means. (This is *not* the same as the mean of the sample variances, which we have just calculated, in the same way that your mother's father is not the same person as your father's mother.) In Example 1 we found that

$$\sigma^2 = 1.5 \quad \text{(see Table 17.1)}$$
$$\sigma_{\bar{x}}^2 = 0.75 \quad \text{(see Table 17.3)}$$

It was no coincidence, in fact, that the variance was halved for the means $\bar{x}$ of samples size 2 compared to the original xs, but we shall understand why and state the exact connection after referring to a more real-life example.

Imagine that you have just landed in England from Mars and as a magnificent 4 ft-tall specimen of Martian personhood would like to know what is the typical height of adult Englishmen. (Incidentally, ponder the surprisingly difficult problem of defining the latter.) Anyway, the first one that you meet happens to be 5 ft 7 in.; but you realise that this may be a poor guide to the mean height of the population, so you (somehow) select at random 4 men from the population (which of course has approximately a normal distribution) and calculate their mean height. You may get

$$\left.\begin{array}{l} 5 \text{ ft } 8 \text{ in.} \\ 5 \text{ ft } 9 \text{ in.} \\ 5 \text{ ft } 10 \text{ in.} \\ 6 \text{ ft } 1 \text{ in.} \end{array}\right\} \text{Mean} = 5 \text{ ft } 10 \text{ in.}$$

This is not likely to be exactly the population mean; but because you have a balancing effect of tall and short men, it is, in general, likely to be closer to it than an individual height. In order to have a sample mean of, say, 5 ft 7 in. the sample would have to contain all short men or, for each one taller than the mean, another one even shorter. If you continue to take samples size 4 and find the mean, these sample means will form a *sampling distribution* that clusters more closely than the original individuals do around an overall (or grand) mean that is equal to the population mean.

This is just as we found in Example 1 for the numbers 2, 2, 3 and 5.

Notice that with people the sampling is strictly without replacement, but from a very large population this is not important.

What happens to this balancing effect with samples larger still?

If we took means of samples of 100 men at a time, these means would cluster even more closely about the population mean, although the effect would diminish. The ultimate would be a sample (the only one possible) consisting of the entire population, whose mean would be, by definition, the population mean and whose p.d.f. would be a single vertical line at the population mean.

The actual connection between the means and variances is

$$\mu_{\bar{x}} = \mu$$

$$\sigma_{\bar{x}}^2 = \frac{\sigma^2}{n}$$

That is,

Mean of sample means = Population mean

Variance of sample means = $\dfrac{\text{Population (i.e. individuals') variance}}{\text{Sample size}}$

To avoid confusion this variance of sample means $\sigma_{\bar{x}}^2$, although it is really just the variance of another distribution, is called the square of the *standard error of the mean.*

Actually, if we know that

$$x \sim N(\mu, \sigma^2)$$

(which is shorthand for 'variable x is normally distributed with mean μ and variance σ^2'), then

$$x \sim N\left(\mu, \frac{\sigma^2}{n}\right)$$

However, it does not matter whether or not the original population was normally distributed (we had one example of each), as the same type of result holds true. Formally stated, the result is known as the *central limit theorem:*

If x has a distribution (not necessarily normal) with mean μ and standard deviation σ, means of samples size n have a distribution with mean μ and standard deviation σ/n, which tends towards a normal distribution shape as n tends towards infinity.

We usually take $n > 30$ as sufficiently large to justify a normal approximation. *Note:*

Standard deviation = $\sqrt{\text{Variance}}$ (by definition)

Figure 17.1 shows the sampling distributions for a typical value of

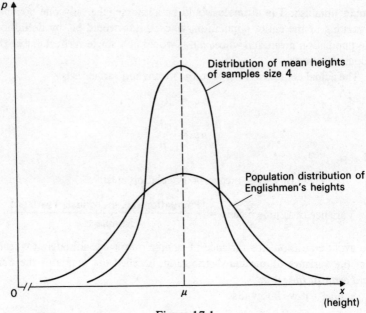

Figure 17.1

n. Since both distributions are p.d.f.s, the areas underneath must be unity, so a tighter cluster implies a taller peak. Thus, for $n = 4$,

$$\text{Standard error } \sigma_{\bar{x}} = \frac{\sigma}{\sqrt{4}} = \frac{\sigma}{2}$$

and the spread is halved; if $n = 100$,

$$\text{Standard error } \sigma_{\bar{x}} = \frac{\sigma}{\sqrt{100}} = \frac{\sigma}{10}$$

This effect achieved with a sample mean is extremely important in other work (e.g. in quality control, because by measuring a mean dimension of a sample of the product we can make much more accurate comments and therefore predictions).

Don't forget that, once obtained, a sample means distribution that is, or can be assumed to be, normally distributed follows the same pattern as any other normal distribution such as we investigated in Chapter 15. A full example will demonstrate this.

Example 2
Samples of size 16 are taken from a normally distributed population with mean 100 and standard deviation 8:

(a) Find the proportion of sample means:

   (i) that exceed 104, 101 and 97 respectively.
   (ii) that lie between 96 and 104.

(b) If a sample of 16 were found to have a mean of 106, what conclusion might you draw?

Since we are given

$$\mu = 100 \quad \sigma = 8 \quad n = 16$$

For the sample distribution,

$$\mu_{\bar{x}} = 100$$

$$\sigma_{\bar{x}} = \frac{\sigma}{\sqrt{n}}$$

$$= \frac{8}{\sqrt{16}} = \tfrac{8}{4} = 2$$

We have the distribution in Figure 17.2.

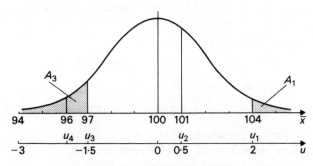

Figure 17.2 Sample means distribution

(a) First, we standardise, using

$$u = \frac{\bar{x} - \mu_{\bar{x}}}{\sigma_{\bar{x}}}$$

(i) Then 104 standardises to

$$u_1 = \frac{104 - 100}{2} = 2 \text{ standard deviations}$$

From table,

$$A_1 = 0.02275 \quad \text{(i.e. 2.275\%)}$$

That is, a proportion of 2.275% of means exceed 104.

Note: We can simply use

$$u = \frac{x - \mu}{\sigma}$$

once we realise that everything stems from a distribution of means.

Similarly, 101 standardises to

$$u_2 = \frac{101 - 100}{2} = 0.5 \text{ standard deviations}$$

From table,

$$A_2 = 0.3085 \quad \text{(i.e. 30.85\%)}$$

That is, *a proportion of 30.85% of means exceed 101*.

We are on the negative side of the standardised normal curve for $\bar{x} = 97$, which standardises to

$$u_3 = \frac{97 - 100}{2} = -1.5 \text{ standard deviations}$$

From table,

$$A_3 = 0.0668 \quad \text{(symmetry)}$$

$$\therefore \text{Area above } u_3 = 1 - A_3$$

$$= 1 - 0.0668 = 0.9332 \quad \text{(i.e. 93.32\%)}$$

That is, a proportion of 93.32% of means exceed 97.

(ii) The range from 96 to 104 represents from -2 to $+2$ standard deviations. From (a)(i) we can see that

$$\text{Area between } u_4 \text{ and } u_1 = 1 - 2A_1$$

$$= 1 - 2(0.02275) = 0.9545 \quad \text{(i.e. 95.45\%)}$$

That is, *a proportion of 95.45% of means lie between 96 and 104*.

(b) A sample mean of 106 standardises to

$$\frac{106 - 100}{2} = 3 \text{ standard deviations}$$

The probability of a value as large as this or larger happening purely by chance is, from the table, only 0.00135 or 0.135%. This is small enough for us to surmise that the sample may not have been taken randomly or that something may have altered in the original population. We shall investigate things like this more precisely in subsequent chapters.

VARIANCE OF SAMPLE MEANS (FINITE POPULATION)

In Example 1 a sample size 2 was taken from 4 numbers *with replacement*. This meant that the first was put back before the second dip was made to complete the sample, so any one of the 4 numbers was available to be chosen again. The *effect* of replacement is therefore as if we were selecting 2 numbers *without* replacement from an infinite (and equally divided) amount of the 4 numbers. We had the 4 numbers 2, 2, 3 and 5 and calculated the mean and variance of this original (infinite) population as $\mu = 3$ and $\sigma^2 = 1.5$ respectively. We also calculated the mean and variance of the sampling distribution of means of samples size 2 as $\mu_{\bar{x}} = 3$ (the same as μ) and $\sigma_{\bar{x}}^2 = 0.75$ ($= \sigma^2/n$ where $n = 2$) respectively. We called $\sigma_{\bar{x}}$ the standard error of the mean.

What if we do *not* replace the first number drawn before dipping for the second? The three numbers left depend on the one already drawn, and the second draw possibilities vary. We do not have each number always available as if from an infinite amount, and we are therefore now sampling from a **finite population**. Let us work through the calculations as before for samples size 2 and see if the result is different for $\mu_{\bar{x}}$ and $\sigma_{\bar{x}}$.

In Table 17.6 the (theoretical) table of all possibilities is constructed as before (compare Table 17.2); but each number cannot be followed by itself, so the diagonal boxes are blank.

Table 17.6 *Sample means $\bar{x}$*

		Second draw			
		2	2	3	5
First draw	2	—	2	2.5	3.5
	2	2	—	2.5	3.5
	3	2.5	2.5	—	4
	5	3.5	3.5	4	—

Hence, we can calculate, in Table 17.7, the mean and variance of the sample means (compare Table 17.3).

We thus get a different result for $\sigma_{\bar{x}}^2$, which is not equal to σ^2/n. Not surprisingly, there is, however, a similar formula for this situation; the

Table 17.7 *Sample means: calculation of mean and variance*

$\bar{x}$	f	$f\bar{x}$	$\bar{x} - \mu_{\bar{x}}$	$(\bar{x} - \mu_{\bar{x}})^2$	$f(\bar{x} - \mu_{\bar{x}})^2$
2	2	4	−1	1	2
2.5	4	10	−0.5	0.25	1
3.5	4	14	0.5	0.25	1
4	2	8	1	1	2
	12	36			6

$$\mu_{\bar{x}} = \frac{\Sigma f\bar{x}}{\Sigma f}$$

$$= \frac{36}{12} = 3$$

$$\sigma_{\bar{x}}^2 = \frac{\Sigma f(\bar{x} - \mu_{\bar{x}})^2}{\Sigma f}$$

$$= \frac{6}{12} = 0.5$$

standard deviation of the sampling distribution of means of samples size n from a finite population of size N (i.e. no replacement) is

$$\sigma_{\bar{x}} = \frac{\sigma}{\sqrt{n}} \sqrt{\frac{N-n}{N-1}}$$

The previous formula $\sigma/\sqrt{n}$ for the standard error of the mean, for samples size n drawn without replacement from an infinite population, is modified by multiplying by $\sqrt{[(N-n)/(N-1)]}$. This is known as the *correction factor for a finite population*. If the sample (size) equals the population, $n = N$ and $\sigma_{\bar{x}} = 0$. If $n = 1$, a sample size 1 is being taken, and $\sigma_{\bar{x}} = \sigma$. The value of the correction factor is often close to 1 when N is large relative to n, so $\sigma_{\bar{x}}$ is approximately equal to $\sigma/\sqrt{n}$. N infinite of course gives $\sigma_{\bar{x}} = \sigma/\sqrt{n}$.

In our example

$$\sigma^2 = 1.5$$

Number in the sample $n = 2$

Number in the population $N = 4$

$$\sigma_{\bar{x}}^2 = \frac{\sigma^2}{n} \left(\frac{N-n}{N-1}\right)$$

$$= \frac{1.5}{2} \left(\frac{4-2}{4-1}\right) = \frac{1.5}{2} \cdot \frac{2}{3} = 0.5$$

which is what we found in Table 17.7.

Remember, finally, that this modified result for $\sigma_{\bar{x}}$ is for sampling without replacement from a finite population. Sampling with replacement from a finite population is equivalent to sampling without replacement from an infinite population.

EXERCISES

(1) Check the results in the chapter for the variances of every possible sample in Example 1, using both the biased and unbiased formulae.

(2) A population consists of 5 members: 3, 7, 7, 8 and 10. Consider all possible samples of size 2 that can be drawn from this population (with replacement). Find the mean and standard deviation of the population, and the mean and standard error of the sampling distribution of the means.

(3) Consider the population of all sample means of samples of 9 drawn from a normally distributed population with $\mu = 100$ and $\sigma = 3$:

(a) What proportion of these 'means':

   (i) exceed 101?
   (ii) lie between 98 and 100?

(b) If a sample of 9 items is drawn and its mean is 104, what is your conclusion?

(4) Samples of size 36 are drawn from a population with mean 100 and standard deviation 9:

(a) What proportion of sample means will exceed:

   (i) 103?
   (ii) 102?

(b) What proportion of sample means will lie between:

   (i) 97 and 103?
   (ii) 98.5 and 101.5?

(c) Within what limits will the central 90% of sample means lie?

(5) Samples size n are drawn from a normally distributed population with standard deviation 5 units. What must n be in order to have

a 99% probability that a sample mean lies within 2 units (either side) of the population mean?

(6) It is known that a refinery produces bags with a mean weight of 80 kg and a standard deviation of 2.6 kg. Find the probability that a random sample of 100 bags will have a mean weight of:
   (a) between 79.4 and 80.6 kg.
   (b) more than 80.8 kg.

18 Confidence and Further Distributions

CONFIDENCE INTERVALS AND LEVELS

We can now formally name something that we have actually met earlier on and again in the exercises of Chapter 17, namely a *confidence interval* (CI).

To take a commonly used confidence interval, we know that for a normally distributed variable 95% of its values lie within 1.96 standard deviations of the mean (see Figure 18.1). In standard units this confidence interval is from -1.96 to $+1.96$, or in original units from $(\mu - 1.96\sigma)$ to $(\mu + 1.96\sigma)$. This is more neatly expressed as the interval $(\mu \pm 1.96\sigma)$ or, generally,

$$\mu \pm u_c \sigma$$

where u_c is the number of standard deviations corresponding to the percentage (i.e. *confidence level*) chosen.

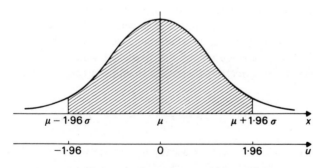

Figure 18.1 Original population distribution

The word 'interval' has its obvious meaning, but the phrase 'confidence interval' is justified because a natural consequence of knowing that 95% of values lie within a certain range is that, when selecting an individual member, we can be 95% sure that its value will be in that range. Recall our definition of a probability.

Although a symmetrical interval is convenient here for the chosen level of confidence, we shall later meet confidence intervals that are not symmetrical.

By altering our level of confidence (i.e. just how many per cent sure we wish to be), we merely change the number of standard deviations u_c. For example,

$$\text{a 99\% CI is } (\mu \pm 2.58\sigma)$$

which means that we are 99% certain that a randomly selected member of a normal population, with mean μ and standard deviation σ, will lie between $(\mu - 2.58\sigma)$ and $(\mu + 2.58\sigma)$.

Using our recently-developed knowledge of sample means, we can say further that

$$\text{if} \quad x \sim N(\mu, \sigma^2)$$

$$\text{then} \quad \bar{x} \sim N\left(\mu, \frac{\sigma^2}{n}\right)$$

for sample size n; so we are 95% confident that a sample mean $\bar{x}$ lies between $(\mu - 1.96\sigma/\sqrt{n})$ and $(\mu + 1.96\sigma/\sqrt{n})$, or

$$\text{a 95\% CI for } \bar{x} \text{ is } \left(\mu \pm 1.96 \frac{\sigma}{\sqrt{n}}\right)$$

Consider just such a sample mean, shown in Figure 18.2 as ∗. Its actual value is unimportant; but if it is within $\pm 1.96\sigma/\sqrt{n}$ of μ, then, by putting the same-width bracket round ∗, we see that, conversely, μ must lie within $\pm 1.96\sigma/\sqrt{n}$ of this particular value of $\bar{x}$. This is true for *any such* $\bar{x}$, not forgetting that 5% of them lie outside;

$$\therefore \text{ a 95\% CI for } \mu \text{ is } \left(\bar{x} \pm 1.96 \frac{\sigma}{\sqrt{n}}\right)$$

This is very important, because usually in practice we do not know μ but can calculate a sample mean $\bar{x}$, which we can then use to make a confidence statement about μ.

Very often, of course, we do not know σ either, and we have already discussed in Chapter 17 how to improve the sample variance s as an estimator, which is particularly important when n is small. We saw how

$$s^2 = \frac{\Sigma(x - \bar{x})^2}{n}$$

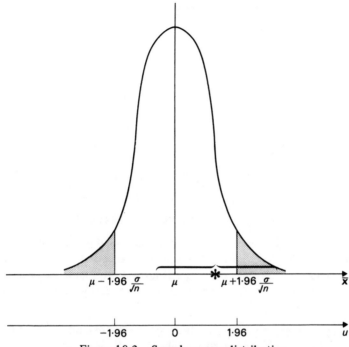

Figure 18.2 Sample means distribution

is biased but $[\Sigma(x-\bar{x})^2]/(n-1)$ is unbiased. We shall use the notation

$$\hat{\sigma}^2 = \frac{\Sigma(x-\bar{x})^2}{n-1}$$

for this unbiased estimator, where $\hat{\sigma}$ (i.e. 'sigma hat') implies that we are estimating σ.

Note that

$$(n-1)\hat{\sigma}^2 = ns^2 \qquad [= \Sigma(x-\bar{x})^2]$$

$$\hat{\sigma}^2 = \frac{ns^2}{n-1}$$

Thus, $\hat{\sigma}$ and s differ merely by a factor.

Example 1

A sample of 64 men, from a population with known standard deviation of height of 2.4 in., gives a mean height of 69.8 in. Find a 90% confidence interval for μ, the mean height of men in the population.

We have $\bar{x} = 69.8$ and $\sigma = 2.4$. Therefore,

$$90\% \text{ CI for } \mu = \bar{x} \pm 1.64 \frac{\sigma}{\sqrt{n}}$$

$$= 69.8 \pm 1.64 \left(\frac{2.4}{\sqrt{64}}\right)$$

$$= 69.8 \pm 1.64 \left(\frac{2.4}{8}\right)$$

$$= 69.8 \pm 1.64 \, (0.3)$$

$$= 69.8 \pm 0.492$$

$$= 69.308 \text{ to } 70.292 \text{ in.}$$

Note: If we had not known that $\sigma = 2.4$ in., we would probably have been told, or been able to calculate, the sample variance s as a similar figure. With a sample that is this large there is no real problem in using such a value for s in place of σ. Later, we shall see how to deal with small samples when σ is unknown.

This is an appropriate point to mention the Procedure Summary Sheets at the end of the book. As we go steadily through the various tests, these sheets summarise what we have done in such a way that, faced with an exercise, you can refer to them and progress logically through a procedure, using what you know from the exercise and being aware of what is not known, to arrive at the correct test. This, I hope, will be helpful while you are first learning about so many easily-confused tests. It is also the best method after you have passed an examination and can use the book as a reference to choose a test in your further study or employment.

If you look at the summary sheets now under Mean μ, you will see that we have just used the first entry in the top left-hand corner. Continue to check like this throughout the book and with every exercise.

SUM AND DIFFERENCE DISTRIBUTIONS

Let us consider a situation where two engineering products are to be assembled together. These could be a piston in a cylinder, or a shaft in a bearing. With mass production it is not possible to try several pistons before obtaining a good fit in a cylinder, so most of the cylinders need to match acceptably with most of the pistons.

What we are really doing is taking a member from a certain population and mixing it (in this case it is the *difference* in dimension that makes the matching acceptable or not) with a member from another population. Assuming that each of these populations are normally distributed, what can we deduce about the likelihood of an acceptable match?

We use the fact that, if we have two distributions of variables x and y such that

$$x \sim N(\mu_x, \sigma_x^2)$$
$$\text{and} \quad y \sim N(\mu_y, \sigma_y^2)$$
$$\text{then} \quad x - y \sim N(\mu_x - \mu_y, \sigma_x^2 + \sigma_y^2)$$

That is, the distribution of the difference between the values of an individual selected from one population and an individual selected from another population has mean equal to the difference of the two population means and variance equal to the sum of the population variances.

Both of these results should become understandable from the preamble to the following example, which also comes up with some typical quantifiable results.

Example 2

If bearings are manufactured with a mean diameter of 50.015 mm and standard deviation 0.006 mm, and shafts have a mean diameter of 50 mm and standard deviation of 0.005 mm, what percentage of randomly chosen bearings and shafts will match, if the acceptable gap is between 0.005 mm and 0.030 mm?

Schematically, we see in Figure 18.3 the situation with a bearing of diameter $x = 50.014$ mm and a shaft of diameter $y = 50.006$ mm. The gap is 0.008 mm, which is acceptable. Why does the acceptable gap not start at 0.000 mm? (Think of the system in operation.)

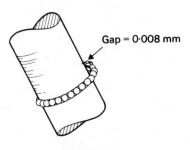

Figure 18.3

If μ is the mean of the gaps (i.e. difference) distribution, it should be clear that it will be the difference between μ_x and μ_y, because every gap is the difference between an x and a y. All the possible values of the gaps will vary around the mean difference of, in this case, 0.015 mm. Thus, we have

$$\mu = \mu_x - \mu_y = 0.015 \text{ mm}$$

If σ is the standard deviation of the gaps distribution, it will not be smaller than σ_x or σ_y, because the possibilities of the *variation* in gap size are larger. Undersize on one variable with oversize on the other, *and vice-versa*, gives a greater variance than for each of the original two populations. In fact,

$$\sigma^2 = \sigma_x^2 + \sigma_y^2$$

Thus, we have

$$\sigma^2 = (0.006)^2 + (0.005)^2$$
$$= 0.000036 + 0.000025$$
$$= 0.000061$$
$$\therefore \sigma = \sqrt{0.000061}$$
$$= 0.00781 \text{ mm}$$

Assuming that all three distributions are approximately normally distributed, we get Figure 18.4.

By standardising as usual, using

$$u = \frac{g - \mu}{\sigma}$$

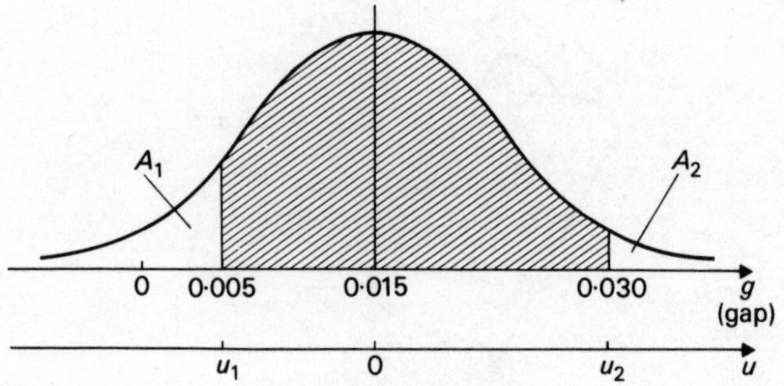

Figure 18.4

we get the lower tolerance u_1 as

$$u_1 = \frac{0.005 - 0.015}{0.00781}$$

$$= -\frac{0.010}{0.00781}$$

$$= -\frac{100}{78.1}$$

$$= -1.28$$

and the upper tolerance u_2 as

$$u_2 = \frac{0.030 - 0.015}{0.00781}$$

$$= \frac{0.015}{0.00781}$$

$$= \frac{150}{78.1}$$

$$= 1.92$$

From tables, we get

$$A_1 = 0.1003 \quad \text{(i.e. 10.03\%)}$$

$$A_2 = 0.0274 \quad \text{(i.e. 2.74\%)}$$

∴ Total unacceptable percentage = 12.77%

∴ Acceptable percentage = 87.23%

You should see that the minus value on the g (gap) scale, and indeed every g value below 0, represents an $(x - y)$ gap that is negative; that is,

$$x - y < 0$$
$$\therefore \quad x < y$$

which means that bearing diameter x is less than shaft diameter y. This is not just below the acceptable tolerance; it represents a physical impossibility. This may of course occur.

Another example of a difference distribution would be matching the heights of men and women. A comparable situation to Example 2 above would be a randomly selected couple having the woman taller than the man (see Exercise 2).

A similar result to that for difference distributions holds for the sum of two (or more) distributions. An example is a product which in manufacture has to pass through workshop A, journey across the factory yard and then pass through workshop B. If the times of its processing, journey and further processing are (for obvious practical reasons) not constant but form approximately normal distributions, the total time of manufacture has a normal distribution with mean equal to the sum of the means and variance equal to the sum of the variances (for similar reasons to the previous argument for a difference — there is a possibly great variation for total manufacture time, between all being carried out early or all being delayed). Formally,

$$\text{if } x \sim N(\mu_x, \sigma_x^2)$$
$$y \sim N(\mu_y, \sigma_y^2)$$
$$z \sim N(\mu_z, \sigma_z^2)$$
$$\text{then } x + y + z \sim N(\mu_x + \mu_y + \mu_z, \sigma_x^2 + \sigma_y^2 + \sigma_z^2)$$

The result adjusts to be true for two distributions and for more than three (see Exercise 3).

PROPORTIONS

A proportion π of, say, the number of people in a population who regularly use the product Sudso, is really just a probability. The connection with frequency (of users) and total frequency is just that

$$\pi = \text{proportion} = \frac{\text{Frequency of users}}{\text{Total frequency}}$$

Since we are concerned with people either using Sudso or not using it (without further detail), we have a *binomial* situation.

We considered this in Chapter 14, and in Chapter 15 there is an example of the use of the normal distribution to approximate to the binomial. To recap,

$$\text{Mean of binomial} = n\pi$$
$$\text{Standard deviation} = \sqrt{[n\pi(1-\pi)]}$$

where n is the number of trials and π is the probability of a 'success' on one trial. The normal approximation is good enough if $n > 30$.

Now, the results for mean and standard deviation above refer to *frequencies*; if $n = 100$ and $\pi = \frac{1}{5}$, then mean = 20 and standard deviation = 4 (check this!). We can easily divide each result by n to convert from frequencies to *proportions*. Thus,

$$\text{Mean proportion} = \frac{n\pi}{n} = \pi \text{ (as expected)}$$

$$\text{Standard deviation of proportion} = \frac{1}{n}\sqrt{[n\pi(1-\pi)]}$$

$$= \sqrt{\frac{n\pi(1-\pi)}{n^2}}$$

$$= \sqrt{\frac{\pi(1-\pi)}{n}}$$

This can be summarised as

$$\left. \begin{aligned} \mu_\pi &= \pi \\ \sigma_\pi &= \sqrt{\frac{\pi(1-\pi)}{n}} \end{aligned} \right\}$$

We can consider μ_π and σ_π as parameters of a sampling distribution of samples size n from a population. Each sample has its own proportion of 'successes' (e.g. people in it who use Sudso) p_1, p_2, p_3, and so on, and we obtain a sampling distribution of proportions, which for $n > 30$ is closely normal.

As before for a mean, we can therefore set up a confidence interval for μ_π as

$$p \pm u_c \sqrt{\frac{p(1-p)}{n}}$$

where p is the sample proportion ($= x/n$), n is sample size and x is the number of successes in the sample. The liberty of using p in calculating the standard deviation is justified by the sample sizes being large.

We can also state now the result for the *difference of proportions* in two populations. (This is to complete the scene presented in this chapter, but an example of its use will await the development of test procedure in Chapter 19.) If two different populations have proportions of 'success' (of the same event) π_1 and π_2 respectively, the distribution of the difference of the proportions obtained in samples size n and N respectively has

$$\left. \begin{array}{l} \text{Mean} = \pi_1 - \pi_2 \\ \text{Variance} = \dfrac{\pi_1(1-\pi_1)}{n} + \dfrac{\pi_2(1-\pi_2)}{N} \end{array} \right\}$$

An example would be the opinion difference between men and women on some political issue (e.g. voting YES in the Common Market referendum). If the samples are large enough, the usual approximation by the normal distribution holds.

(A similar result for the sum of proportions is not quoted here because it is not very useful and might confuse.)

Finally, we shall use Example 3 to illustrate the *probable error* of an estimate.

Example 3

It is known from past experience that the standard deviation of the reaction time in a test for road safety is 0.04 seconds. If an experimenter wants to be 95% confident that his error in estimating the mean reaction time does not exceed 0.01 seconds, how large a sample must he take?

In a confidence interval for μ we had

$$95\% \text{ CI is } \left(\bar{x} \pm u_c \frac{\sigma}{\sqrt{n}} \right)$$

which is our estimate of μ, expressed in probability terms over an interval. The width that we allow each side of $\bar{x}$ (in this case $u_c \cdot \sigma/\sqrt{n}$) is the probable error of the estimate. The actual value of $\bar{x}$ is not needed for our purpose here.

We have

$$u_c \frac{\sigma}{\sqrt{n}} = \text{Probable error}$$

$$\therefore 1.96 \left(\frac{0.04}{\sqrt{n}} \right) \leqslant 0.01$$

by substituting and making sure that the probable error is less than or equal to the required limit.

$$\therefore 1.96 \times 4 \leqslant \sqrt{n} \times 1$$
$$7.84 \leqslant \sqrt{n}$$
$$(7.84)^2 \leqslant n$$
$$61.47 \leqslant n$$

That is, the experimenter needs n to be 62 people (or more) for the confidence required.

Similarly, the error in the estimate of a proportion is

$$u_c \sqrt{\frac{p(1-p)}{n}}$$

If sample size n is the only unknown, simple arithmetic by substituting known values for the other variables will yield it.

EXERCISES

(1) A random sample of 150 grummets made by a particular grummet machine in a day were weighed and found to have a mean weight of 0.102 kg with a standard deviation of 0.005 kg. Calculate (a) 95% and (b) 99% confidence intervals for the mean weight of grummets.

(2) If the heights in inches of men and women are known to be distributed as $N(68, 2^2)$ and $N(64, 2^2)$ respectively, what is the probability that a randomly chosen couple has the woman taller than the man?

(3) During its assembly a mass-produced product has to pass through Departments A and B of a factory (with negligible transit time between them), then be transferred before passing through department C. The table shows the mean and standard deviation of the processing times (in minutes) in each stage of assembly. Assuming that all distributions are normal, find the mean and standard deviation of the complete assembly-time distribution.

Calculate the time under which 95% of products are assembled.

Stage	Mean	Standard deviation
Department A	35	3
Department B	20	2.1
B to C	3	0.7
Department C	22	2

(4) A sample poll of 100 voters chosen at random in the country town of Little-Doing-in-the-Daytime contained 45% who supported the

candidate of the Leftovers Party. Find 90% confidence limits for the proportion of all voters in the town who supported this candidate.

(5) The Inflato Tyre Company wishes to determine the average mileage of its new PQ tyres. A random sample of 49 tyres gives a mean of 25,000 miles with a standard deviation of 2,450 miles:

   (a) Find a 90% confidence interval for the true mean mileage (i.e. of all such tyres).
   (b) If we kept the same sample (and therefore size), with what degree of confidence could we claim that the error of our estimate was within ± 500 miles?
   (c) How large a sample would be necessary to claim with 90% confidence that the error of our estimate is within ± 500 miles?

(6) A sample of 100 part-time students at a polytechnic showed that 25 dropped out in the first year of study:

   (a) Calculate a 95% confidence interval for the drop-out rate of the population of all comparable students.
   (b) What sample size is necessary to give the confidence interval in (a) a precision of ± 5%?

19 Hypothesis Testing

ONE-TAIL AND TWO-TAIL TESTS

In the last chapter our confidence intervals gave us a range within which we can have a certain probability of finding a population mean or a population proportion from sample information.

Such estimation is useful, but suppose now that we want to tighten up the whole procedure and test whether a mean or a proportion is what we hypothesise (i.e. believe or claim) that it is. A statistical hypothesis is usually of the *null* (i.e. neutral) type.

Example 1
If we believe that a machine is possibly more efficient under a new operative, we set up a **null hypothesis** that its mean hourly output is unchanged. The **alternative hypothesis** in this case is that the mean hourly output is actually improved. Using the sample information that is available we can then test the hypothesis.

Let us suppose that, from experience, we know the mean output to be 427 widgets per hour, with a standard deviation of 24 widgets. The new operative achieves a mean hourly output of 440 widgets over 16 one-hour runs.

We call the null hypothesis H_0 and the alternative hypothesis H_A (or H_1), so we have

H_0: $\mu = 427$ (i.e. mean hourly output unchanged)

H_A: $\mu > 427$ (i.e. mean hourly output improved)

Remember that the population of all hourly outputs achieved, which we assume closely follows a normal distribution, contains freak figures that may be due solely to chance and not to any real improvement. Our test is going to decide on the balance of probabilities what seems to be happening. A picture of the situation will underline this (see Figure 19.1), remembering also that it is the sampling distribution of means of samples size 16 from which our figure of 440 has come.

By taking the mean of 16 one-hour runs, instead of just one run, we

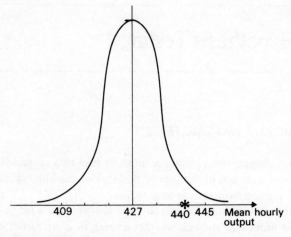

Figure 19.1 Distribution of sample means ($n = 16$)

tighten up the spread of values on which we are going to decide, by a factor of 4. Thus,

$$\text{Standard error} = \frac{\sigma}{\sqrt{n}}$$

$$= \frac{24}{\sqrt{16}} = \frac{24}{4} = 6$$

What we are really testing is: 'Is this *mean score* of 440 due to a chance extreme value or to a genuine improvement?' Our decision will necessarily be made in probability terms and may therefore be wrong.

We have the information

$\bar{x} = 440$ (i.e. new operative's mean score)

$\mu = 427$ (assumed true by H_0)

$\frac{\sigma}{\sqrt{n}} = 6$ (σ and n given)

We shall therefore test $\bar{x}$ by standardising it, in exactly the same way as we have often done before, as

$$\frac{\text{Value of variable} - \text{Mean of variable}}{\text{Standard deviation of variable}}$$

in order to give a number of standard deviations representing how extreme the value of the variable is. Since we can assume that the original population was normal and that its standard deviation was known, we are testing with the statistic u, where

$$u = \frac{\bar{x} - \mu}{\frac{\sigma}{\sqrt{n}}}$$

Substituting, we get

$$\text{Numerator} = 440 - 427 = 13$$

$$\text{Denominator} = \frac{24}{\sqrt{16}} = \frac{24}{4} = 6$$

$$\therefore u = \frac{13}{6} = 2.17$$

Our decision is made on an arbitrary level of confidence (or *significance*, which is the complement of it), strictly speaking chosen before calculating the statistic u, but we can take the commonly used level of 95%. This means that we compare our u value of 2.17 with the figure from normal tables of the number of standard deviations below which 95% of all values lie. Table 4 in Murdoch and Barnes* shows this to be 1.645 (see Figure 19.2). The level that we take depends on the context; a test on the effectiveness of a new drug might need a very high level, for example.

Our value of 2.17 exceeds 1.645, a result that would only happen by chance 5% of the time, so we conclude that the chances are 19 to 1

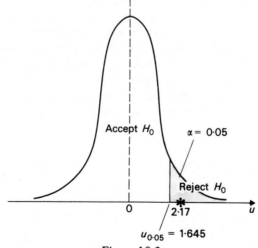

Figure 19.2

*See Neave, page 20.

that it has not happened by chance or that 'the result is significant at the 5% level'.

We reject the null hypothesis H_0 and say that it seems (at this level of significance) that the new operative really is better.

If we had taken the 1% significance level, our u statistic would have been compared (see Table 4 again) with 2.3263, so the result would not have been significant and H_0 would have been accepted; 2.17 is not extreme enough to have been beyond what happens purely by chance only 1% of the time.

Because we set up H_A by saying that any change was strictly for the better, we have what is called a *one-tail test*, a term that follows from Figure 19.2, which is the pictorial expression of the consequences of H_0 and H_A.

An hypothesis H_A set up simply as H_0 being untrue (e.g. any change may be higher *or* lower) would involve, at the 5% level of significance, two equal tails of $2\frac{1}{2}$%. This is known as a *two-tail test* and is demonstrated in Example 2 in a test for proportion, worked fully but without the introductory detail of the last example.

Example 2

A firm runs an aptitude course during which it is believed from past experience, that 40% of applicants will be found unsuitable. A batch of 54 applicants take the course, and 16 turn out to be unsuitable. Is this result significantly different?

First, I should mention that it is tempting for students to make every test, and especially those for proportions, into a one-tail test, if only because the sample figure is bound to be either up or down on what is expected. Be careful to apply a one-tail test *only* when you have some outside or collateral reason to believe a change is really in a particular direction. If in doubt, use a two-tail test.

Note also that we assume that the new batch is a random sample. A rejected H_0 could indicate suspicions about the figures rather than some real change.

We have

H_0: $\pi = 0.4$ (i.e. the population proportion)

H_A: $\pi \neq 0.4$ (i.e. simply different)

We take a level of significance of $\alpha = 0.05$.
We have a two-tail test.
We use the statistic

$$u = \frac{p - \pi}{\sqrt{\frac{\pi(1-\pi)}{n}}}$$

since $n > 30$, and p is the sample proportion of a distribution that is therefore closely normal, with

$$\text{Mean} = \pi$$

$$\text{Standard deviation} = \sqrt{\frac{\pi(1-\pi)}{n}}$$

We are really just standardising as for $\bar{x}$ before.

Now,

$$p = \tfrac{16}{54} = 0.296$$

Substituting in our test statistic u, we get

$$\text{Numerator} = 0.296 - 0.4$$
$$= -0.104 \quad \text{(3 decimal places)}$$

$$\text{Denominator} = \sqrt{\frac{0.4(1-0.4)}{54}}$$

$$= \sqrt{\frac{0.24}{54}}$$

$$= \sqrt{\tfrac{1}{225}}$$

$$= \tfrac{1}{15}$$

$$\therefore u = \frac{-0.104}{\tfrac{1}{15}} = -0.104 \times 15 = -1.56$$

The area for rejection (sometimes referred to as the *critical region*) is shown in Figure 19.3, where the critical values of $u = \pm 1.96$ come from the table for $\alpha = 0.025$. Now, -1.56 is not in the critical region, so we accept H_0. It seems that the batch is not significantly different from the norm.

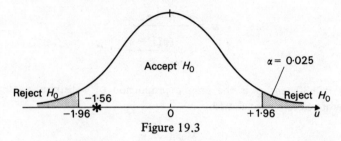

Figure 19.3

TYPE I AND II ERRORS

Although we choose (in Example 2 at the 0.05 level) to say that a u value in the critical region is significant and H_0 is rejected because only 1 time in 20 will such a value have occurred by chance, nevertheless this 1 in 20 (which is actually the area α) is the probability that we reject the null hypothesis when it is in fact true. This 'α risk' is called a *type one error*. For example, compare the chance of finding a defendant guilty when he is actually innocent, because judicial decisions are also made on a balance of probabilities and perhaps there is freak circumstantial evidence.

Not quite so easy to understand mathematically is the *type two error* (or 'β risk'), which occurs when we fail to reject a false null hypothesis and it is accepted as true. For example, compare a guilty man's being found innocent, perhaps through lack of evidence.

In our Example 1 a type I error would be to conclude that the new operative is better when he is really about the same or perhaps even worse, as we have a one-tail test (i.e. to reject a true H_0). A type II error would be to conclude that he is no better when really he is (i.e. to accept a false H_0).

The balance between the two types of error could be important in the relative freedom of a judicial system or in deciding about a new drug, for example (see Exercise 3). To conclude wrongly that a new drug is not better (a type II error) means that it will not be applied and that people may suffer or die unnecessarily. On the other hand, applying a new drug through wrongly thinking it is better (a type I error) will, at least, waste lots of money in its production, if it is no better; if it is actually a worse drug (a one-tail test), it may even be dangerous.

The decision on α is obviously a value-judgement over *which* error is the more serious. If α is very small (e.g. a very high standard is required to consider a drug as better), we may never change.

The section that follows can be left and read separately if you wish to continue on to the next test, which is for the difference of means.

CALCULATION OF β (THE PROBABILITY OF A TYPE II ERROR) AND OPERATING CHARACTERISTIC CURVES

When a null hypothesis is tested, we have first to set up the null hypothesis and then a criterion by which to test it. Since the null hypothesis can be true *or* false but the decision is made to accept or reject it on the basis of satisfying the criterion (which can possibly happen even with a false hypothesis), there are four situations that can exist. They can be neatly summarised as in Table 19.1.

Table 19.1

	Criterion says: Accept H_0	Reject H_0
H_0 is true	Correct	Type I error
H_0 is false	Type II error	Correct

For convenience we can use the figures of our Example 1 but modify the alternative hypothesis to a two-tail test. Thus, we set up

H_0: $\mu = 427$ (i.e. the mean hourly output of operative)

H_A: $\mu \neq 427$ (i.e. operative better or worse)

We might choose as the criterion (equivalent to a significance level) that the operative's mean hourly output is accepted as 427 if a sample of 16 one-hour runs gives a mean output of between 415 and 439. We know that the standard deviation for all hourly runs is $\sigma = 24$, and thus the standard error (i.e. the standard deviation of means of samples size 16) is

$$\text{Standard error} = \frac{\sigma}{\sqrt{n}} = \frac{24}{\sqrt{16}} = 6$$

Our criterion of between 415 and 439 means 12 either side of 427 or, in standardised form,

$$u = \frac{\text{Gap}}{\text{Standard error}}$$
$$= \pm \frac{12}{6} = \pm 2$$

which is the equivalent criterion if H_0 is true (see Figure 19.4).

Thus, *if* H_0 *is true* and the mean is 427, the chance of rejecting H_0 (a type I error) is the tail(s) area beyond $u = \pm 2$, which from tables is $\alpha = 0.0455$. In other words, the probability α of obtaining a value outside

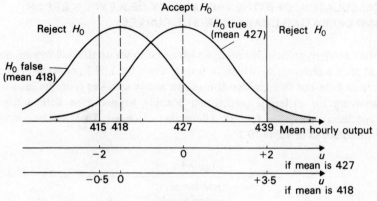

Figure 19.4

the range 415 to 439 (by chance, since H_0 is true) and committing the type I error of rejecting a true null hypothesis is 0.0455 in this case.

On the other hand, *if* H_0 *is not true*, we cannot calculate the probability β of falsely accepting it unless the value (which is *not* 427) of the mean is known. H_A contains all alternatives to H_0 so we would have to calculate β for every possible value of the mean, other than 427. See Figure 19.4 again, where the possibility of the mean's being 418 (which shifts the distribution) is checked against the test criterion. The limits of the acceptance range now standardise to

$$\frac{415-418}{6} \quad \text{and} \quad \frac{439-418}{6}$$

i.e. -0.5 and $+3.5$

A type II error is committed when H_0 is falsely accepted, and this occurs, with a mean of 418, when the criterion is met by a value in the (now standardised) range -0.5 to $+3.5$. From tables, the chance of this is given by the area $(0.1915 + 0.49977)$. Thus, the probability β of committing a type II error because a value between 415 and 439 has occurred and H_0 is accepted, even though the mean is 418, is 0.69127.

Remember that α is really determined by setting the test criterion, so a change in α will alter the criterion and therefore alter β. Nonetheless, for each α there will be a different β for every different value of the mean that is possible. In practice, several β values are calculated and plotted against the mean value that gave rise to them, in what is called the *operating characteristic* curve or OC curve. You should see easily from Figure 19.4 that, as the value of the mean gets further away from 427 (i.e. the value under H_0), so the distribution shifts over and β gets smaller. The probability of 0.69127 that we believe the mean to be 427 when it is really 418 seems high, but the discrepancy in mean value is small. The probability of believing the mean to be 427 when it is really, say, 409 is very much smaller (see Figure 19.5).

415 relative to 409 standardises to $+1$, so the chance of accepting H_0 is given by the area 0.1587 from tables. Therefore, if the mean is 409, $\beta = 0.1587$.

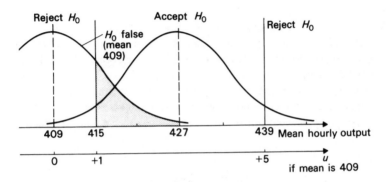

Figure 19.5

As an *exercise* calculate β for the mean values

$$403, 406, (409), 412, 415, (418), 421 \text{ and } 424$$

and their equivalents on the right hand side

$$430, 433, 436, 439, 442, 445, 448 \text{ and } 451$$

Plot each β against the mean that produced it to find the OC curve. Note that the OC curve defined as plotting β against mean value would have no entry at 427, because by definition H_0 true means that the error of falsely accepting it is impossible. We therefore actually plot 'probability of accepting H_0' against 'mean value'; everywhere except at 427 this is the same as β, but at 427 it is just $(1-\alpha)$, and this fills the gap.

The probability of (correctly) rejecting a false hypothesis is $(1-\beta)$, and this is called the *power of the test*. Plotting $(1-\beta)$ against the mean values gives the inverse of the OC curve, known as the *power curve*. Can you describe the value needed at 427, which is actually α?

Finally, let us consider what the probabilities of the two types of error may mean in an industrial situation where a producer supplies a product at a mean hypothesised value to a consumer. The *producer's risk* α is that a consignment that is really okay gets sent back after a sample produces a rejection. The *consumer's risk* β is that a consignment that is actually off the desired mean value gets used after a sample produces acceptance.

A change in the test criterion or in the sample size will alter α and β, so the OC curve can be made to have any desired shape.

TEST FOR DIFFERENCE OF MEANS

We shall now extend our previous work on difference distributions in Chapter 18 to distributions of means. This is not really new at all, since a sample-means normal distribution is simply a special case of an individual's normal distribution.

Instead of subtracting two individual values taken one from each of two different populations, we are now comparing sample means $\bar{x}$ and $\bar{y}$ from two sampling distributions of means of samples size n and N respectively. Formally,

$$\text{if} \quad \bar{x} \sim N\left(\mu_x, \frac{\sigma_x^2}{n}\right) \quad \text{(samples size } n\text{)}$$

$$\text{and} \quad \bar{y} \sim N\left(\mu_y, \frac{\sigma_y^2}{N}\right) \quad \text{(samples size } N\text{)}$$

$$\text{then} \quad \bar{x} - \bar{y} \sim N\left(\mu_x - \mu_y, \frac{\sigma_x^2}{n} + \frac{\sigma_y^2}{N}\right)$$

Example 3

Two comparable makes of light bulb, the Brightabulb (B) and the Supershine (S), are tested to compare their lifetimes. A sample of 100 of B had a mean life of 1,018 hours with a standard deviation of 50 hours, while a sample of 120 of S had a mean life of 1,009 hours with a standard deviation of 60 hours. Is there any significant difference in the lifetime of each make?

We are considering the mean of each sample, say of 100 of B, as one member of a complete distribution of all the means of samples size 100 that could be drawn from the original population of B bulbs; similarly with the S bulbs.

Then, the difference between a mean from the B means and a mean from the S means does itself have a distribution, with

$$\text{Mean} = \mu_x - \mu_y$$

$$\text{Standard deviation} = \sqrt{\left(\frac{\sigma_x^2}{n} + \frac{\sigma_y^2}{N}\right)}$$

Strictly, this distribution is normal only if we draw from normal distributions (which are independent). As before, however, a large-enough sample size always gives a resulting distribution that can be considered as approximately normal and allows us to use sample standard deviations as good estimators for σ_x and σ_y.

We shall test with a statistic that is again just a particular case of our 'standardising' formula

$$\frac{\text{Value of variable} - \text{Mean of variable}}{\text{Standard deviation of variable}}$$

namely (assuming normality),

$$u = \frac{(\bar{x} - \bar{y}) - (\mu_x - \mu_y)}{\sqrt{\left(\frac{\sigma_x^2}{n} + \frac{\sigma_y^2}{N}\right)}}$$

Our null hypothesis will be that there is no difference in mean lifetimes, with the alternative that there is just some difference (i.e. a two-tail test):

H_0: $\mu_x = \mu_y$ (i.e. equal mean lifetimes)
H_A: $\mu_x \neq \mu_y$ (i.e. different mean lifetimes)

We can take the (more extreme) 0.01 level of significance for variety, although this example is not a case of a life-or-death decision.

We have

$$\bar{x} = 1{,}018 \qquad \bar{y} = 1{,}009$$
$$s_x = 50 \qquad s_y = 60$$
$$n = 100 \qquad N = 120$$

We take s_x and s_y as good estimators for σ_x and σ_y, since n and N are large.

Substituting in our test statistic u, we get

$$\text{Numerator} = (1{,}018 - 1{,}009) - (0) = 9$$

since $\mu_x - \mu_y = 0$ on the assumption of H_0 true, and

$$\text{Denominator} = \sqrt{\left(\frac{2{,}500}{100} + \frac{3{,}600}{120}\right)}$$

$$= \sqrt{(25 + 30)}$$

$$= \sqrt{55}$$

$$= 7.42$$

$$\therefore u = \frac{+9}{7.42} = +1.21$$

At the 0.01 level (two-tail) we compare with $+2.58$ (see Figure 19.6) and have a non-significant result. We accept H_0. It seems that there is no real difference in the mean lifetimes of the Brightabulb and the Supershine.

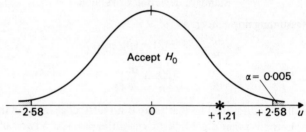

Figure 19.6

TEST FOR DIFFERENCE OF PROPORTIONS

We have already mentioned in Chapter 18 the distribution of the difference of proportions, and stated its parameters, but it is now convenient to revise and demonstrate this by referring to Example 4 below.

Example 4

A survey is commissioned by the Government of Ruritania to test whether men and women have different views on capital punishment, by asking the question: 'Are you in favour of capital punishment?' In a sample of 800 men there were 536 replies of 'Yes', and in a sample of 700 women there were 427 replies of 'Yes'. By ignoring the problem of 'don't knows', decide at the 0.10 level if there is a significant difference in attitude due to sex.

Hypothesis Testing

Imagine that we continually draw samples size n from the population of all men, which has a true proportion of π_1, saying 'Yes'. The sample proportions will vary around π_1 with variance $\pi_1(1-\pi_1)/n$. Similarly, samples size N from all the women, who have a true proportion π_2 of replies 'Yes', will have sample proportions varying around π_2 with variance $\pi_2(1-\pi_2)/N$.

If we take a (size n) sample proportion p_1 of the men and subtract a (size N) sample proportion p_2 of the women, repeatedly doing this gives us a sampling distribution of proportions difference, with

$$\left. \begin{array}{l} \text{Mean} = \pi_1 - \pi_2 \\ \text{Variance} = \dfrac{\pi_1(1-\pi_1)}{n} + \dfrac{\pi_2(1-\pi_2)}{N} \end{array} \right\}$$

If n and N are large, the distribution is approximately normal.

We test the hypothesis that the men and women have no difference in attitude; that is, they reply with equal proportions of 'Yes', as if from the same population of people with a certain attitude. Hence,

H_0: $\quad \pi_1 = \pi_2 = \pi \quad$ (i.e. proportions equal)

H_A: $\quad \pi_1 > \pi_2 \quad$ (i.e. men more likely to reply 'Yes')

Using our usual 'standardising' rule for the test statistic, we should test with

$$u = \frac{(p_1 - p_2) - (\pi_1 - \pi_2)}{\sqrt{\left(\dfrac{\pi_1(1-\pi_1)}{n} + \dfrac{\pi_2(1-\pi_2)}{N}\right)}}$$

Unfortunately, we do not know π_1 or π_2; but as we are assuming that they are equal under H_0, the statistic at least contracts to

$$u = \frac{(p_1 - p_2) - (0)}{\sqrt{\left(\dfrac{\pi_1(1-\pi_1)}{n} + \dfrac{\pi_1(1-\pi_1)}{N}\right)}}$$

using π_1, but π_2 would do equally well as a symbol. Therefore, using π, it is

$$u = \frac{p_1 - p_2}{\sqrt{\left[\pi(1-\pi)\left(\dfrac{1}{n} + \dfrac{1}{N}\right)\right]}}$$

We can estimate π (which is unknown) from the samples by pooling (i.e. the weighted arithmetic mean); thus,

$$\text{Estimate } \pi = \frac{np_1 + Np_2}{n + N}$$

We have

$$p_1 = \tfrac{536}{800} = 0.67 \qquad p_2 = \tfrac{427}{700} = 0.61$$

$$n = 800 \qquad N = 700$$

$$\therefore \text{Estimate } \pi = \frac{536 + 427}{800 + 700} = \frac{963}{1,500}$$

Substituting in our test statistic u, we get

$$\text{Numerator} = 0.67 - 0.61 = 0.06$$

$$\text{Denominator} = \sqrt{\left[\frac{963}{1,500}\left(1 - \frac{963}{1,500}\right)\left(\frac{1}{800} + \frac{1}{700}\right)\right]}$$

$$= \left[0.642\,(0.358)\left(\frac{15}{5,600}\right)\right]$$

$$= \sqrt{0.0006156}$$

$$= 0.0248$$

$$\therefore u = \frac{0.06}{0.0248} = 2.419$$

At the 0.10 level we compare with 1.28 and find the result to be significant (see Figure 19.7). We reject H_0. It seems that men are more in favour of capital punishment than women.

Look carefully at the test statistic used in Example 4. Note that, if the *same* sample proportions p_1 and p_2 had been obtained from larger samples,

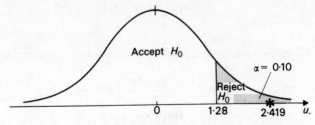

Figure 19.7

this would have made the denominator of the statistic *smaller* and thus the result *larger*. Consequently, a given difference between two sample proportions is more significant if the samples are large than if they are small, which is what you should expect intuitively.

EXERCISES

(1) The workers' representative in the building industry in the north produces figures from a sample of 400, which show an average weekly wage of £71.4 with a standard deviation of £6.2. Can he claim that this proves northern workers to be lower paid than the national building-industry average wage of £72.5?

Choose your own significance level, and explain your test fully, stating any assumptions.

(2) A very large firm believes that the absence record of its office employees is 4.3 days/employee/year. A sample of the absences of 100 office employees over the past year shows the information in the table. Find the mean and standard deviation of this sample, and use them to test whether there has been a change in the absence record at the 0.05 level of significance. State clearly your assumptions.

Number of days' absence	Frequency
0	4
1	6
2	11
3	17
4	27
5	17
6	8
7	5
8	2
9	1
10	2
More than 10	0

(3) A medical research team uses a certain 'improved' drug on 50 newly-admitted patients suffering from a serious illness, and 11 of them are

cured. If it is known that previously the cure rate was only 10%, is the team justified in believing that the new drug is more effective?

(4) The Mobex petrol company wants to show that its petrol (with the miracle ingredient Eekout) gives greater miles per gallon than the petrol of rival company Yesno. A test with 60 cars on Mobex gives a mean m.p.g. of 28.4 with standard deviation of 2.2, and 50 cars (of comparable type, condition, and so on) run on Yesno give a mean of 27.7 m.p.g. with standard deviation 1.9 m.p.g. Can a claim be made for significantly better performance at the 0.10 level?

(5) The manufacturer of Biobub washing powder has a market survey carried out in two regions to determine the consumer preference for his product. In region A a sample of 80 people contains 44 people who prefer Biobub to all other powders, and in region B a sample of 60 people contains 26 people for whom it is the preferred powder. Should the manufacturer believe that there is a significant difference in the regional proportions who prefer Biobub?

20 Small Samples and the t Distribution

THE STUDENT t DISTRIBUTION

We have so far carried out only tests where the population standard deviation(s) or population proportion(s) was known *or* where the samples taken were *large enough* to allow us to estimate them from the sample standard deviation(s) or sample proportion(s). Either way this gave us the justification for using the normal distribution on our test statistic, whatever it was.

What happens, however, if we do not know σ and the sample size is *not* large enough to justify using s as an estimate and then testing with the normal distribution? The point is that for a small n (i.e. <30) we simply cannot assume that the sampling distribution is normal, when we do not know σ.

Intuitively, you may think that we can apply the same principles in our test as before but, because of a small sample size n, we need to have less confidence in accepting or rejecting the hypothesis. The smaller n is, the less confidence we have; and the greater n becomes, the more our confidence grows, until the previous levels with the normal distribution are reached.

This is exactly what we do, using a distribution known as the **Student t distribution**. It is symmetrical and looks rather like a flattened normal distribution with larger tails. These larger tails fulfil our intuitive requirement of having to go further out from the centre in order to get the same central area (or same tail area) as in the normal distribution (see Figure 20.1).

The t distribution has the same basic shape for all values of n but gets nearer and nearer to the normal distribution shape as n increases. Figure 20.1 shows the comparison for $n = 12$.

A tail area of $\alpha = 0.05$ gives (from Murdoch and Barnes's Table 4* of the normal distribution) $u_\alpha = 1.645$. The same tail area of $\alpha = 0.05$ gives (from Murdoch and Barnes's Table 7† of the t distribution) $t_{\alpha;\nu} = 1.796$ for $\nu = 11$.

*See Neave, page 18.
†See Neave, page 20.

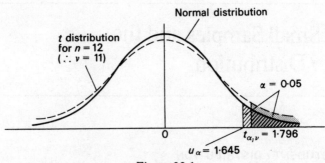

Figure 20.1

This new Greek symbol ν (i.e. nu, the Greek n) is directly related to the sample size n and expresses what are called the *degrees of freedom*. The number of degrees of freedom ν shows how restrictions affect the sample size n. Consider the analogy of a ruler that is free to move in three dimensions until pinned at one point to a table and consequently only free to move (i.e. pivot) in two dimensions. We have one restriction, which leads to

$$\nu = n - 1$$

In other contexts the connection between ν and n is slightly different. We shall also meet a further explanation of the idea of 'degrees of freedom' in the next chapter.

Run your eye down the 0.05 column of Table 7 to see how the t values approach the u values as ν increases (i.e. as n increases).

We shall now work through some examples where n is small and σ is unknown, so the t distribution has to be used. Because n is small, we shall also (at last) use our unbiased estimator $\hat{\sigma}$, instead of the sample standard deviation s, to estimate an unknown σ.

CONFIDENCE INTERVALS

Example 1
A factory produces castings whose weights are assumed to be distributed normally. A sample of 10 castings has weights (in kilograms) as follows:

7.4   7.9   8.2   8.7   6.9   7.2   7.7   8.4   8.7   8.9

Find the mean and standard deviation weight of the sample, and use it to provide a 95% confidence interval for the true mean weight of all castings.

First, we calculate $\bar{x}$ and s or $\hat{\sigma}$ (see Table 20.1).

Table 20.1 *Calculation of mean and standard deviation (n = 10)*

x	$x - \bar{x}$	$(x - \bar{x})^2$
7.4	−0.6	0.36
7.9	−0.1	0.01
8.2	0.2	0.04
8.7	0.7	0.49
6.9	−1.1	1.21
7.2	−0.8	0.64
7.7	−0.3	0.09
8.4	0.4	0.16
8.7	0.7	0.49
8.9	0.9	0.81
80.0		4.30

$$\bar{x} = \frac{\Sigma x}{n}$$

$$= \frac{80}{10} = 8$$

$$\hat{\sigma} = \sqrt{\frac{\Sigma(x - \bar{x})^2}{n - 1}}$$

$$= \sqrt{\frac{4.30}{9}} = \sqrt{0.478} = 0.691$$

Note: If $\bar{x}$ is not a convenient round figure, use the alternative formula for variance:

$$s^2 = \frac{\Sigma x^2}{n} - (\bar{x})^2$$

We have previously used a confidence interval for μ of

$$\bar{x} \pm u_c \cdot \frac{\sigma}{\sqrt{n}}$$

We now use t instead of u and $\hat{\sigma}$ instead of σ, since n is small and σ is unknown; our confidence interval for μ becomes

$$\bar{x} \pm t_c \cdot \frac{\hat{\sigma}}{\sqrt{n}}$$

where $\nu = n - 1$ and t_c (or $t_{\alpha;\nu}$) depends on ν and the α chosen.

From Table 7 we get

$$t_{0.025;9} = 2.262 \qquad (\nu = 9, \quad \alpha = 0.025 \text{ for each tail})$$

Our CI is thus

$$8 \pm 2.262 \left(\frac{0.691}{\sqrt{10}}\right) = 8 \pm 0.494$$

That is, we are 95% sure that the true mean weight μ lies between 7.51 and 8.49 kg.

TEST FOR MEANS

Example 2

The Industrial Placement Unit of Unisex Polytechnic believes that the average salary paid to sandwich students during their industrial year is £2,800. A sample of 17 of its own students reveals that their average salary is £2,860 with a standard deviation of £105. Does this evidence suggest that the countrywide average salary is higher than £2,800?

We are given that

$$\bar{x} = £2,860$$
$$s = £\ 105$$
$$n = £\ 17$$

If we had been given the sample as 17 individual salary figures, we could of course have calculated $\bar{x}$ and s (or $\hat{\sigma}$), just as we did in Example 1. If an exercise that you have to do gives the 'sample variance' but does not state clearly whether it is s or $\hat{\sigma}$, either ask for further clarification from your tutor or, failing that, assume that it is s.

Since the sample is clearly small and σ is unknown, we need to test the information with the t distribution.

We need to change s to $\hat{\sigma}$, but the connection is just

$$\frac{\hat{\sigma}}{\sqrt{n}} = \frac{s}{\sqrt{(n-1)}}$$

The test statistic that we shall use is almost the same as the one that we used before for a test on a mean when we were justified in using the normal distribution, namely

$$u = \frac{\bar{x} - \mu}{\frac{\sigma}{\sqrt{n}}}$$

We simply replace σ by $\hat{\sigma}$ and u (i.e. the normal distribution statistic) by t (i.e. the t distribution statistic); that is,

$$t = \frac{\bar{x} - \mu}{\frac{\hat{\sigma}}{\sqrt{n}}}$$

Because of my remarks above about sometimes being given $\hat{\sigma}$ or s in an exercise, you may see tutors use a denominator of $s/\sqrt{(n-1)}$ instead of $\hat{\sigma}/\sqrt{n}$. My own preference is to use $\hat{\sigma}$, because it preserves the same form that we use in all our tests, and that is the main purpose of this book: if you can see that all tests are really just one general principle applied to several particular cases, it should be much easier to understand. Most students think (wrongly) that every situation has its own test that has no relation to any other.

Thus, we have

$$H_0: \quad \mu = 2{,}800$$
$$H_A: \quad \mu > 2{,}800 \quad \text{(i.e. a one-tail test)}$$

Although our own sample mean $\bar{x}$ is higher, we have no actual collateral information to suggest that a one-tail test is the only appropriate one. A two-tail test could be used.

Substituting in our test statistic t, we get

$$\text{Numerator} = 2{,}860 - 2{,}800 = 60$$

$$\text{Denominator} = \frac{105}{\sqrt{16}} = \tfrac{105}{4}$$

$$\therefore t = 60 \cdot \tfrac{4}{105} = \tfrac{16}{7} = 2.286$$

Degrees of freedom are

$$\nu = n - 1 = 16$$

We can take a suitable level of significance of 0.05, which for a one-tail test means that the tail is 0.05. From Table 7, looking across to $\alpha = 0.05$ and down to $\nu = 16$, we get

$$t_{0.05;16} = 1.746$$

Our result exceeds this and so is in the critical region (see Figure 20.2). We therefore reject H_0.

The result is significant at the 0.05 level, although notice from the table that $t_{0.01;16} = 2.583$, and so the result would not be significant at the 0.01 level.

The conclusion that we draw from the significance depends on our assumptions. We have assumed that the sample is a random one from all comparable students (i.e. the population, which is assumed to have approximately a normal distribution, of industrial year salaries). If our sample is random, rejecting H_0 means that the true mean salary appears to be higher than £2,800. As in many practical cases the sample, although random from the polytechnic, may be in a region that is able to obtain consistently higher salaries (e.g. London) and is not really random from the whole population.

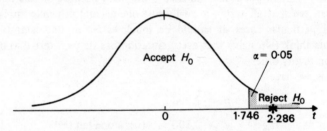

Figure 20.2 t distribution ($\nu = 16$)

TEST FOR DIFFERENCE OF MEANS

Example 3

Two different brands of matchboxes (costing the same) are compared according to the average number of sticks that they contain. A random sample of 8 boxes of brand A gives the following figures:

$$46 \quad 45 \quad 48 \quad 44 \quad 47 \quad 44 \quad 48 \quad 46$$

A random sample of 10 boxes of brand B gives the following figures:

$$43 \quad 44 \quad 46 \quad 45 \quad 45 \quad 46 \quad 44 \quad 46 \quad 44 \quad 47$$

(a) Test if the brands differ in mean contents.
(b) If A claims to give more for its money, decide on the claim.

Show clearly each step in the construction and testing of your hypothesis, and state any assumptions that you make.

We want to compare the brands by testing the difference of their mean contents. This is what we did in Chapter 19 to compare the mean lifetime of two types of light bulb; but now we are not given the standard deviation of either distribution of contents, and the samples are too small to justify testing a statistic with the normal distribution.

We shall have to adapt our test using the t distribution, and first we need to calculate $\bar{x}$ and s_x (for brand A) and $\bar{y}$ and s_y (for brand B) from the samples, size n and N respectively (see Table 20.2).

Table 20.2 *Calculation of means and estimated variance ($n = 8$, $N = 10$)*

x	$x - \bar{x}$	$(x - \bar{x})^2$	y	$y - \bar{y}$	$(y - \bar{y})^2$
46	0	0	43	-2	4
45	-1	1	44	-1	1
48	2	4	46	1	1
44	-2	4	45	0	0
47	1	1	45	0	0
44	-2	4	46	1	1
48	2	4	44	-1	1
46	0	0	46	1	1
368		18	44	-1	1
			47	2	4
			450		14

$$\bar{x} = \frac{\Sigma x}{n} \qquad \bar{y} = \frac{\Sigma y}{N}$$

$$= \frac{368}{8} = 46 \qquad = \frac{450}{10} = 45$$

$$\hat{\sigma}^2 = \frac{\Sigma(x - \bar{x})^2 + \Sigma(y - \bar{y})^2}{n + N - 2}$$

$$= \frac{18 + 14}{8 + 10 - 2} = \frac{32}{16} = 2$$

The test statistic that we used when the normal distribution was appropriate was

$$u = \frac{(\bar{x} - \bar{y}) - (\mu_x - \mu_y)}{\sqrt{\left(\frac{\sigma_x^2}{n} + \frac{\sigma_y^2}{N}\right)}}$$

for samples size n and N respectively.

We are faced with the problem of how to estimate σ_x and σ_y. We have to assume that, *even though they are unknown, they are equal.* The reason for this is that, if the standard deviations are both unknown and unequal, we are in mathematically deep water beyond the scope of this book. In fact we meet later a test for variances to see whether this assumption is valid in just such a situation as this, but for now we assume that

$$\sigma_x = \sigma_y \,(=\sigma)$$

The test statistic then contracts to

$$u = \frac{(\bar{x} - \bar{y}) - (\mu_x - \mu_y)}{\sqrt{\left[\sigma^2 \left(\frac{1}{n} + \frac{1}{N}\right)\right]}}$$

Even so, we still do not know what they are equal to, so we make an estimate for σ using s_x and s_y. We take the weighted (by sample size) arithmetic mean of the sample variances s_x^2 and s_y^2 to give the estimate $\hat{\sigma}^2$ for σ^2. This is a very similar method to our estimate for p in the test for difference of sample proportions. Notice the adjustment in the denominator, however, which is similar to our previous adjustment to produce an unbiased estimator. Hence,

$$\hat{\sigma}^2 = \frac{ns_x^2 + Ns_y^2}{n + N - 2}$$

The $(n + N - 2)$ can be considered as made up of an $(n - 1)$ for the x distribution and an $(N - 1)$ for the y distribution.

(Notice the alternative forms for $\hat{\sigma}^2$ when s_x^2 and s_y^2 have to be calculated from the sample, namely

$$\hat{\sigma}^2 = \frac{\Sigma(x - \bar{x})^2 + \Sigma(y - \bar{y})^2}{n + N - 2}$$

which is used in Table 20.2 and is equivalent to

$$\hat{\sigma}^2 = \frac{[\Sigma x^2 - n(\bar{x})^2] + [\Sigma y^2 - N(\bar{y})^2]}{n + N - 2}$$

Check your previous variance formulae to be sure of these.)

Our test statistic becomes

$$t = \frac{(\bar{x} - \bar{y}) - (\mu_x - \mu_y)}{\sqrt{\left[\hat{\sigma}^2 \left(\frac{1}{n} + \frac{1}{N}\right)\right]}}$$

We have

(a) H_0: $\mu_x = \mu_y$ (b) H_0: $\mu_x = \mu_y$
H_A: $\mu_x \neq \mu_y$ H_A: $\mu_x > \mu_y$
(i.e. a two-tail test)   (i.e. a one-tail test)

Substituting in our test statistic t, we get

$$\text{Numerator} = (46-45) - (0) = 1$$

$$\text{Denominator} = \sqrt{\left[2\left(\frac{1}{8}+\frac{1}{10}\right)\right]} = \sqrt{\left(2 \times \frac{9}{40}\right)} = \frac{3}{4.472}$$

$$\therefore t = 1 \times \frac{4.472}{3} = 1.491$$

The degrees of freedom are

$$\nu = n + N - 2$$
$$= 8 + 10 - 2 = 16$$

The reasoning for ν is that now we have two restrictions. We choose the 0.05 level of significance. Looking in Table 7, we get:

(a) $t_{0.025;16} = 2.12$ (b) $t_{0.05;16} = 1.746$

$\therefore$ 1.491 is not significant $\therefore$ 1.491 is not significant

$\therefore$ We accept H_0 $\therefore$ We accept H_0

That is, it seems at this level of significance (a) that there is no difference between the mean contents of the two brands of matchbox, and also (b) that brand A cannot claim to give more matchsticks.

PAIRED SAMPLES

It sometimes happens that we have what appear to be two different sets of sample figures, when really they refer to the same 'before and after' people (or objects). Each person (or object) then has two figures that can be paired together (e.g. weight before and after a diet). We must always have a genuine reason for the pairing.

What really interests us in this situation is the difference between each pair of figures for every person (e.g. either a gain or loss in weight). The situation thus reduces to one (paired) sample, where the figures are the set of differences in weight for every person.

The differences d will form a (usually small) sample with its own mean $\bar{d}$ and its own sample variance s_d. The sample can be considered as coming from the population of weight differences of all people on the same diet conditions, and under a null hypothesis the mean weight difference μ_d should be zero. The variance σ_d^2 is usually unknown.

With these conditions a test using the t distribution can be carried out, just as in Example 2.

Example 4

The Weightwobblers Club uses 8 of its members to evaluate a new 14-day diet. They are weighed before the diet and again a fortnight after using it, as shown by the results in Table 20.3. Has there been a loss in weight because of the diet?

Table 20.3

Member	A	B	C	D	E	F	G	H
Weight before (kg)	61.4	64.0	59.8	62.6	63.2	61.7	63.2	65.3
Weight after (kg)	59.2	62.7	58.1	61.4	61.3	63.9	60.4	66.2

For convenience we subtract each after-weight from each before-weight (to give losses). The d values are assumed to be from a population that is normal, at least approximately (see Table 20.4).

We have

H_0: $\quad \mu_d = 0 \quad$ (i.e. mean difference zero, diet no effect)

H_A: $\quad \mu_d > 0 \quad$ (i.e. a real loss in weight)

We could of course have the differences the other way round and obtain the other tail in this one-tail test. (We assume that the diet can only, on the face of it, cause loss.)

We test with

$$t = \frac{\bar{d} - \mu_d}{\frac{\hat{\sigma}_d}{\sqrt{n}}}$$

since the sample is small and σ_d is unknown.

Substituting in the test statistic t, we get

$$\text{Numerator} = 1 - (0) = 1$$

$$\text{Denominator} = \frac{1.69}{\sqrt{8}} = \frac{1.69}{2.828}$$

$$\therefore t = 1 \times \frac{2.828}{1.69} = 1.67$$

The degrees of freedom are

$$\nu = n - 1$$

$$= 8 - 1 = 7$$

At the 0.05 level of significance we get, from Table 7,

$$t_{0.05;7} = 1.895 \qquad (>1.67)$$

$$\therefore 1.67 \text{ is not significant}$$

$$\therefore \text{ We accept } H_0$$

It seems that the diet has no significant effect.

Table 20.4 *Calculation of mean and estimated standard deviation (n = 8)*

d	$d - \bar{d}$	$(d - \bar{d})^2$
2.2	1.2	1.44
1.3	0.3	0.09
1.7	0.7	0.49
1.2	0.2	0.04
1.9	0.9	0.81
−2.2	−3.2	10.24
2.8	1.8	3.24
−0.9	−1.9	3.61
8.0		19.96

$$\bar{d} = \frac{\Sigma d}{n}$$

$$= \frac{8.0}{8} = 1$$

$$\hat{\sigma}_d^2 = \frac{\Sigma(d - \bar{d})^2}{n - 1}$$

$$= \frac{19.96}{8 - 1} = \frac{19.96}{7} = 2.85$$

$$\therefore \hat{\sigma}_d = 1.69$$

EXERCISES

(1) A travel company takes a random sample of 16 similar families who took package holidays and finds that the mean cost of their holidays was £320 with a standard deviation of £30. Construct a 98% confidence interval for the mean cost of one of their package holidays.

(2) The Sparko company claims that its batteries have an average life of 3 years. The owner of a small taxi fleet of 6 cars uses Sparko batteries and from his records discovers that they have the following lifetimes (in months):

$$29 \quad 38 \quad 31 \quad 35 \quad 32 \quad 33$$

Use these figures to test the Sparko claim at the 0.05 level of significance.

(3) An intelligence test is carried out in a college on 12 students randomly chosen from the BA Inhumanities Degree, and they obtain a mean score of 110 points with a standard deviation of 5 points. A random sample of 12 students from the BSc Bureaucracy Degree scores a mean of 117 points with a standard deviation of 7 points. Does this suggest a significant difference in students' intelligence between the two degrees?

(4) A random sample is taken of 4 mines from the North and 6 mines from the South. Their productivity, measured on the appropriate index of production with 1978 = 100, is given in the table. Test these figures to see if there is any significant difference between the productivity of North and South in the mining industry.

North	103	106	103	102		
South	106	107	105	108	103	107

Show clearly the construction of your test, explaining the reasons behind any assumptions that you make.

HINT: To calculate the sample variances, subtract 100 from each figure and use the alternative variance formula.

(5) A company selects 9 salesmen at random, and their sales figures for the previous month are recorded. They then undergo a training

course devised by a business consultant, and their sales figures for the following month are compared, as shown in the table. Has the training course caused an improvement in the salesmen's ability?

Previous month	75	90	94	85	100	90	69	70	64
Following month	77	101	93	92	105	88	73	76	68

21 The χ^2 Distribution and Its Uses

THE χ^2 DISTRIBUTION

Consider an experiment of throwing a single die 120 times and recording the number of times on which each individual score appears. We might obtain the result in Table 21.1. If the die is fair, we expect the perfect result to be the same frequency of each score (i.e. 20), but we would not be very surprised at the frequencies tabulated. They vary around the 'expected' number of 20, but not wildly. How can we measure the difference of the whole set from the perfect result?

Table 21.1 *Single die: first 120 throws*

Score	1	2	3	4	5	6
Frequency	16	25	22	20	11	26

Thinking simply, as usual, we try our previous and very successful ideas used in the construction of the formula for variance. If each *observed frequency* from the experiment is referred to by the symbol O and the corresponding *expected frequency* by the symbol E, the deviation from perfection is expressed by $(O-E)$, which can be either positive or negative (see Table 21.2).

To remove problems of sign we square each deviation. This is because it is deviation, not its direction, that interests us. Hence, we get

$$(O-E)^2 \quad 16 \quad 25 \quad 4 \quad 0 \quad 81 \quad 36$$

We now adjust each squared deviation by comparing it with (i.e. dividing to find the ratio) its *own* expected frequency. This is because the deviation

Table 21.2 *Single die: first 120 throws*

Score	1	2	3	4	5	6
Observed, O	16	25	22	20	11	26
Expected, E	20	20	20	20	20	20
$O-E$	-4	5	2	0	-9	6

of 16 from 20 is relatively much more important than the deviation of 996 from 1,000. The ratio $(O-E)^2/E$ captures this factor; for example,

$$\frac{(16-20)^2}{20} = \frac{16}{20} = 0.8$$

whereas $\quad\dfrac{(996-1,000)^2}{1,000} = \dfrac{16}{1,000} = 0.016$

For our experimental figures we get

$$\frac{(O-E)^2}{E} \quad \tfrac{16}{20} \quad \tfrac{25}{20} \quad \tfrac{4}{20} \quad \tfrac{0}{20} \quad \tfrac{81}{20} \quad \tfrac{36}{20}$$

It is important to realise that, although each expected frequency in this case is the same, there will be cases (see later examples) where they are particular to each category (or score).

The next logical step is to add up these individual 'relative squared deviations', to get

$$\Sigma \frac{(O-E)^2}{E} = \tfrac{16}{20} + \tfrac{25}{20} + \tfrac{4}{20} + \tfrac{0}{20} + \tfrac{81}{20} + \tfrac{36}{20}$$

$$= \tfrac{162}{20}$$

$$= 8.1$$

This value of 8.1 measures how different our set of observed frequencies is from the expected frequencies in a very similar way to a variance, although it does so using the concept of 'relative' rather than 'average' on the squared deviations. The actual connection between χ^2 and variance is explored in the next chapter.

The figures that we have used are obviously only one set of the many possible. A further experiment might yield the figures and calculations in Table 21.3. The value of 1.4 obtained for $\Sigma[(O-E)^2/E]$ indicates the closer agreement of the second set of figures.

We could repeat the experiment (of 120 throws) any number of times and calculate the same measure each time. We would get a distribution of numbers varying between zero (when the observed figures were all the same as the 'expected' of 20) and a very large number (when the disagreement was as great as possible). Can you see what the largest possible value of $\Sigma[(O-E)^2/E]$ is in our experiment? Most of the measures, like our values 8.1 and 1.4, would be between these extremes. The overall picture

Table 21.3 *Single die: second 120 throws*

Score	1	2	3	4	5	6
O	18	22	23	19	21	17
E	20	20	20	20	20	20
$O-E$	-2	2	3	-1	1	-3
$\dfrac{(O-E)^2}{E}$	$\tfrac{4}{20}$	$\tfrac{4}{20}$	$\tfrac{9}{20}$	$\tfrac{1}{20}$	$\tfrac{1}{20}$	$\tfrac{9}{20}$

$$\sum \frac{(O-E)^2}{E} = \tfrac{28}{20} = 1.4$$

of the distribution of values follows the shape shown in Figure 21.1. It is known as the χ^2 *distribution*, and thus we can summarise our previous measure as

$$\chi^2 = \sum \frac{(O-E)^2}{E}$$

Notice that the distribution is not symmetrical and depends for its exact shape on the degrees of freedom ν. As before, ν is related to k (i.e. the number of categories or classes) by the restrictions in the situation. If you look back at either set of experimental results, you should see that the frequencies in each category must add up to 120, so knowing any 5 of the 6 frequencies will determine the remaining frequency. This is our restriction. Thus,

Degrees of freedom = Number of categories − Number of restrictions

In this situation

$$\nu = k - 1$$
$$= 6 - 1 = 5$$

The areas in the tail of the χ^2 distribution are given in Table 8 of Murdoch and Barnes[*] for different values of ν. Check that for $\alpha = 0.05$ and $\nu = 5$, we get

$$\chi^2_{0.05;5} = 11.07$$

This means that only 5% of experiments like ours would give a measure of difference $\Sigma [(O-E)^2/E]$ as large (or larger) than 11.07.

Our results of 8.1 and 1.4 were consequently not significantly large, and this should lead you to realise that the χ^2 distribution can be used

[*]See Neave, page 21.

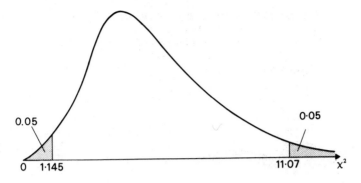

Figure 21.1 χ^2 distribution ($\nu = 5$)

to test the agreement (i.e. what is properly called the *goodness of fit*) of a set of experimental figures to the set of expected figures. The usual testing procedure of setting up a null hypothesis (which will determine the expected frequencies) and an alternative hypothesis, choosing the significance level, and so on can be followed.

Notice that Table 8 also gives 'large' α tails; for example, the left-hand (non-symmetrical) 0.05 tail can be found by looking across to $\alpha = 0.95$ and down to $\nu = 5$, where we find 1.145 (see Figure 21.1). It might be useful to know this when testing, because it would show a measure of agreement only equalled or bettered in 5% of experiments, which might be 'too good to be true'. Our own second experiment gave a χ^2 value of 1.4, which is quite close to 1.145.

Normally, however, we shall be testing a discrepancy between figures.

One small aside is in order before we work a full goodness-of-fit example. When the degrees of freedom are just 1 and the total frequency is not very large, an error is introduced into the calculation of χ^2, which we overcome by using *Yates' correction*. This merely involves subtracting 0.5 from each deviation $(O - E)$ before squaring, and so on.

Suppose that we toss a coin 50 times and get the frequencies shown in Table 21.4. Since $k = 2$ (categories), $\nu = 1$; and if the coin is fair, $E = 25$. So we get a value of 1.3 for χ^2, using Yates' correction. If we are testing at the 0.05 level, Table 8 gives

$$\chi^2_{0.05;1} = 3.841 \quad (\text{and } \chi^2_{0.95;1} = 0.00393)$$

so our experimental result is not significantly different from that expected from a fair coin.

Table 21.4 *Coin tossing: calculation of χ^2 using Yates' correction (n = 50)*

	Heads	Tails
Observed, O	29	21
Expected, E	25	25
$O - E$	4	-4
$O - E - 0.5$	3.5	-4.5
$(O - E - 0.5)^2$	12.25	20.25
$\dfrac{(O - E - 0.5)^2}{E}$	$\dfrac{12.25}{25}$	$\dfrac{20.25}{25}$

$$\chi^2 = \sum \frac{(O - E - 0.5)^2}{E}$$

$$= \frac{12.25}{25} + \frac{20.25}{25} = \frac{32.5}{25} = 1.3$$

GOODNESS OF FIT

Our single die experiment above gave figures that although we did not specifically say so, were tested against the fit of a rectangular distribution where all frequencies were equally likely (see Figure 21.2).

We can test the fit of any distribution using χ^2, and in particular we shall now do examples involving the binomial, Poisson and normal distributions.

It is important to realise that certain conditions are necessary to justify testing the comparison between a discrete distribution (classifying a continuous distribution for practical reasons makes it effectively discrete) and

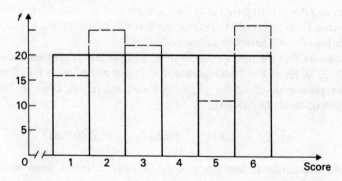

Figure 21.2

The χ^2 Distribution and Its Uses 263

a continuous distribution. We usually consider that a χ^2 test can be used if the total frequency is at least 30 (or, even better, 50) and if each category contains an expected frequency of at least 5 (or, even better, 10). If there are less than 10 (usually in an end category), combination with another category solves this.

Example 1

A study was made of 320 families with 5 children. The distribution of the possible number of boys (B) and girls (G) that they contained is given in Table 21.5. Do these figures support the hypothesis that the probabilities of male and female births are equal?

Table 21.5

Combination	(5B)	(4B, 1G)	(3B, 2G)	(2B, 3G)	(1B, 4G)	(5G)
Observed	17	56	108	89	41	9

We have

$$H_0: \quad p(B) = p(G) \quad \text{(i.e. binomial, } p = \tfrac{1}{2}\text{)}$$
$$H_A: \quad p(B) \neq p(G)$$

Since $p = \tfrac{1}{2}$ under H_0, our previous work on the binomial distribution (just think of the relevant line of Pascal's triangle) gives expected ratios of

$$1:5:10:10:5:1 \quad \text{(Total 32)}$$

Since we have 320 families, each frequency is multiplied by 10 to give expected frequencies E of

$$10 \quad 50 \quad 100 \quad 100 \quad 50 \quad 10 \quad \text{(Total 320)}$$

Using these we can calculate that $\chi^2 = 9.19$ (see Table 21.6). Now, $\nu = 5$, since we have $k = 6$ categories and one restriction on total frequency (p is assumed known, but, if it were not and had to be calculated from a sample, the extra restriction would make ν equal to 4). From Table 8, with $\alpha = 0.05$, we get

$$\chi^2_{0.05;5} = 11.07$$

Our result is therefore not significant, and it seems that there is equal probability of a male or female birth. The binomial distribution with $p = \tfrac{1}{2}$ is accepted as a good fit to the observed data.

Table 21.6 *Calculation of* χ^2

Combination	(5B)	(4B, 1G)	(3B, 2G)	(2B, 3G)	(1B, 4G)	(5G)
O	17	56	108	89	41	9
E	10	50	100	100	50	10
$O - E$	7	6	8	-11	-9	-1
$(O-E)^2$	49	36	64	121	81	1
$\dfrac{(O-E)^2}{E}$	$\dfrac{49}{10}$	$\dfrac{36}{50}$	$\dfrac{64}{100}$	$\dfrac{121}{100}$	$\dfrac{81}{50}$	$\dfrac{1}{10}$

$$\chi^2 = \sum \frac{(O-E)^2}{E}$$

$$= 4.9 + 0.72 + 0.64 + 1.21 + 1.62 + 0.1$$

$$= 9.19$$

Example 2

The Haso Company takes a random sample of 500 of its new radios and records the number of faults in each, as in Table 21.7. Calculate the mean and variance of the sample. Also, test the hypothesis that the number of faults follows a Poisson distribution.

Table 21.7

Number of faults	0	1	2	3	4	5 or more
Frequency	253	176	45	21	4	1

The sample mean and variance are calculated in Table 21.8, taking the last category as 5. Remember that the mean and variance of a Poisson distribution should be equal. The values here are not too disparate.

We try fitting a Poisson distribution with mean $m = 0.7$. From Table 2 (Murdoch and Barnes) of *cumulative* Poisson probabilities[*] we get, for $m = 0.7$, the probabilities $p(r)$ in Table 21.9. Remember that each $p(r)$ for an exact number r is calculated by subtracting $p(r+1$ or more) from $p(r$ or more); for example,

$$p(3) = p(3 \text{ or more}) - p(4 \text{ or more})$$

Next we multiply each probability by 500 to get the expected frequencies shown in Table 21.10.

We have to combine categories to satisfy our conditions for χ^2 which is calculated in Table 21.11.

[*]See Neeve, page 17.

Table 21.8 *Calculation of mean and variance*

x	f	xf	$(x - \bar{x})$	$(x - \bar{x})^2$	$f(x - \bar{x})^2$
0	253	0	−0.7	0.49	123.97
1	176	176	0.3	0.09	15.84
2	45	90	1.3	1.69	76.05
3	21	63	2.3	5.29	111.09
4	4	16	3.3	10.89	43.56
5	1	5	4.3	18.49	18.49
	500	350			389.00

$$\bar{x} = \frac{\Sigma xf}{\Sigma f}$$

$$= \tfrac{350}{500} = 0.7$$

$$s^2 = \frac{\Sigma f(x - \bar{x})^2}{\Sigma f}$$

$$= \tfrac{389}{500} = 0.778$$

Table 21.9

r	$p(r$ or more$)$	$p(r)$
0	1.0000	0.4966
1	0.5034	0.3476
2	0.1558	0.1217
3	0.0341	0.0283
4	0.0058	0.0050
5	0.0008	0.0007
6	0.0001	0.0001
7	0.0000	—
		1.0000

Table 21.10

x	Observed frequency, O	Expected frequency, E
0	253	248.30
1	176	173.80
2	45	60.85
3	21	14.15
4	4	2.50
5	1	0.35
6	0	0.05
	500	500.00

Table 21.11 Calculation of χ^2

x	O	E	O−E	$(O-E)^2$	$\dfrac{(O-E)^2}{E}$
0	253	248.30	4.70	22.09	0.09
1	176	173.80	2.20	4.84	0.03
2	45	60.85	−15.85	251.19	4.13
3 or more	26	17.05	8.95	80.10	4.70
	500	500.00			8.95

$$\chi^2 = \sum \frac{(O-E)^2}{E} = 8.95$$

We are restricted by total frequency and by not knowing the mean m of the Poisson distribution; therefore,

$$\nu = k - 2$$
$$= 4 - 2 = 2$$

With $\alpha = 0.05$, from Table 8 we get

$$\chi^2_{0.05;2} = 5.991$$

so we have a significant result and are not justified in assuming a good fit to the observed faults distribution by the Poisson distribution with $m = 0.7$. The number of faults suggests that they are not independent.

Example 3
Using the figures of Exercise 2, Chapter 19, of absences from employment (given again in Table 21.12), test to see if they are consistent with the assumption of having come from a normal distribution.

Table 21.12

Number of days' absence, x	f
0	4
1	6
2	11
3	17
4	27
5	17
6	8
7	5
8	2
9	1
10	2

You were required to calculate the mean and standard deviation in the previous exercise in order to test a belief about the mean, so we shall assume that you did it and obtained the figures

$$\bar{x} = 4$$
$$s = 2.005$$
$$n = 100$$
$$k = 11 \quad \text{(i.e. categories or classes)}$$

Notice immediately that we have three restrictions here: on the total frequency of 100, plus the unknown μ and σ of the normal distribution that we shall try to fit on to the distribution; hence,

$$\nu = k - 3$$
$$= 11 - 3 = 8$$

The histogram of the distribution is Figure 21.3.

We attempt to fit a normal distribution as shown in Figure 21.4 using the $\bar{x}$ and s calculated and comparing the areas (which are of course probabilities) around each of the possible number of days' absence with the corresponding frequency in the histogram.

Notice that each division under the normal curve (which is continuous) needs to be from 0.5 below the number to 0.5 above. We have met this before during a normal approximation to a discrete distribution. In order to compare directly a probability with a frequency (e.g. p_5 and f_5 shown

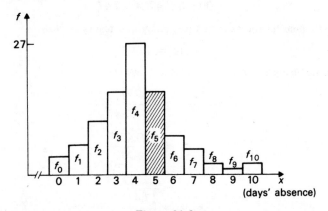

Figure 21.3

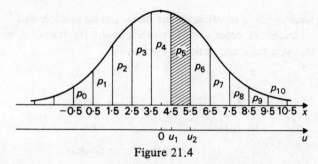

Figure 21.4

shaded) we shall first have to multiply the probability by the total frequency n. We can then compare np_5 (i.e. the expected frequency from the normal curve that we are fitting) with f_5 (i.e. the corresponding observed frequency from the sample).

First, we need to calculate p_5 from tables. Standardising, using

$$u = \frac{x - \bar{x}}{s}$$

$$u_1 = \frac{4.5 - 4}{2.005} = \frac{0.5}{2.005} = 0.25 \quad \text{standard deviations}$$

$$u_2 = \frac{5.5 - 4}{2.005} = \frac{1.5}{2.005} = 0.75 \quad \text{standard deviations}$$

From Table 3, we get

$$p_5 = 0.4013 - 0.2266 = 0.1747$$

$$\therefore E_5 = np_5$$

$$= 100 \times 0.1747 = 17.47$$

where E_5 denotes the expected frequency in category 5. Now,

$$O_5 = 17$$

Therefore, using χ^2 on the fifth category, we get

$$\frac{(O_5 - E_5)^2}{E_5} = \frac{(17 - 17.47)^2}{17.47}$$

$$= \frac{(-0.47)^2}{17.47}$$

$$= \frac{0.2209}{17.47}$$

$$= 0.013$$

Thus, the contribution to χ^2 from the fifth category is 0.013.

There is obviously a fair amount of work involved in calculating the E values for all the other categories, although they are done in exactly the same way. They can then be used to calculate

$$\chi^2 = \sum \frac{(O-E)^2}{E}$$

I leave you to complete this example as Exercise 5. *Hint*: You may combine the bottom two categories and also the top four categories to meet our conditions, thus giving seven over all. This will give

$$\nu = k - 3$$
$$= 7 - 3 = 4$$

Using the χ^2 table with $\nu = 4$ will tell you whether the result is significant or not and therefore whether the fit is bad or good.

CONTINGENCY TABLES

It often happens that we have a variable that can be classified according to two attributes and we are interested in whether these are independent. Sociology uses this situation often; an example would be to classify a sample of voters both by party preference and by region, so that every voter falls into a category relating to one party and one region. If the two criteria of classification are independent (i.e. the null hypothesis), it is possible to calculate what number should be in each category or 'box' of the two-dimensional table by giving it just its fair proportion.

The χ^2 distribution can be used to test such a hypothesis, and we shall explain everything with (as usual) a full worked example.

Example 4

The Providential and Fortuitous Insurance Company uses the type of region that a policy holder's home is in to decide the premium. Table 21.13 gives the figures revealed by records for a random sample of 500 policy holders, who are classified as shown both by region and by claim level. Decide whether there is a connection between claim level and type of region.

First, we calculate the subtotals according to each classification. These can all be seen together in Table 21.14 a little later on; but to demonstrate the calculation of a typical expected frequency, let us consider the

Table 21.13 *Observed number of claimants*

	City	Urban	Rural
Claim £200 or more	21	22	7
Claim under £200	33	58	9
No claim	96	170	84

classifications of 'City', of which there are 150 policy holders (all claim levels), and 'Claim under £200', of which there are 100 policy holders (in all regions).

We have

H_0: No connection between region and claim level

H_A: Some connection

This will be a one-tail test, because we are not likely to be interested in a 'too good to be true' result.

How many *should* there be (i.e. expected) in the box containing the observed 33 from a 'City' region who 'Claim under £200'? Since there are 150 'City' policy holders out of the 500 total, the hypothesis that they are not connected to any particular claim level means that they should represent $\frac{150}{500}$ (i.e. $\frac{3}{10}$) of every claim subtotal. Since there are 100 'Claim under £200' policy holders, the 'City' subsection should be $\frac{3}{10}$ of those 100; that is,

$$\frac{150}{500} \times 100 = 30$$

Similarly, we could arrive at 30 by considering the correct proportion or fair share as being $\frac{1}{5}$ (i.e. 100 'Claim under £200' out of 500 total) of the 150 'City' policy holders.

Every box is worked out in the same way. The central column is the easiest if you are still not clear about where these expected figures come from in Table 21.14. Since 'Urban' is exactly half of the total (i.e. 250 out of 500), it should have half of every claim level, hence 25, 50 and 175 out of 50, 100 and 350 respectively.

Table 21.14 *Expected number of claimants*

	City	Urban	Rural	Total
Claim £200 or more	15	25	10	50
Claim under £200	30	50	20	100
No Claim	105	175	70	350
Total	150	250	100	500

You can remember a little formula to calculate each expected value, namely,

$$\text{Expected value} = \frac{\text{Product of each subtotal}}{\text{Total}}$$

but it is much better to understand what you are doing.

We now have all the figures that we need to do a χ^2 test. Substituting, we get

$$\chi^2 = \sum \frac{(O-E)^2}{E}$$

$$= \frac{(21-15)^2}{15} + \frac{(22-25)^2}{25} + \frac{(7-10)^2}{10} + \frac{(33-30)^2}{30}$$

$$+ \frac{(58-50)^2}{50} + \frac{(9-20)^2}{20} + \frac{(96-105)^2}{105}$$

$$+ \frac{(170-175)^2}{175} + \frac{(84-70)^2}{70}$$

$$= \tfrac{36}{15} + \tfrac{9}{25} + \tfrac{9}{10} + \tfrac{9}{30} + \tfrac{64}{50} + \tfrac{121}{20} + \tfrac{81}{105} + \tfrac{25}{175} + \tfrac{196}{70}$$

$$= 2.4 + 0.36 + 0.9 + 0.3 + 1.28 + 6.05 + 0.77 + 0.14 + 2.8$$

$$= 15$$

We take the 0.05 level of significance as convenient.

The number of degrees of freedom depends on our restriction of subtotals. Looking at either table, you should see that in each row knowing two figures predetermines the third. The same is true for each column. However many rows (or columns) there are, the last figure is always determined by the subtotal. Thus, in general,

$$\nu = (r-1)(c-1)$$

where r is the number of rows and c is the number of columns. In this example we have

$$\nu = (3-1)(3-1) = 2 \times 2 = 4$$

(Try filling in *any* four figures in a 3 by 3 (i.e. nine-box) table, and you will find that the other five are determined. Thus, a 6 by 4 table will have $5 \times 3 = 15$ degrees of freedom, and so on.)

Looking in Table 8, we find

$$\chi^2_{0.05;4} = 9.488$$

Our result of 15 is bigger and therefore lies in the critical region (see Figure 21.5). We have a significant result and reject H_0. It seems that there is a connection between region and claim level.

Just as a footnote, look at Exercise 6 carefully and realise that a test for difference of proportions could be carried out on the information.

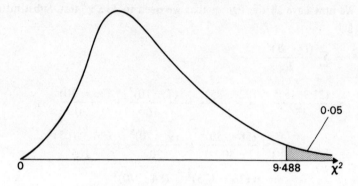

Figure 21.5 χ^2 distribution ($\nu = 4$)

EXERCISES

(1) A coin is tossed 60 times in a test for bias, and 38 heads and 22 tails are recorded. Calculate the χ^2 value for these figures, first without and then with Yates' correction.

(2) Of the 600 books borrowed from a college library during one week, the pattern of the number borrowed each day was as tabulated. Is there a connection between the number of books borrowed and the day of the week?

Day	Monday	Tuesday	Wednesday	Thursday	Friday
Number borrowed	126	131	103	134	106

(3) A check is carried out on the number of errors made during each hour of an 8 hour shift by certain skilled workers in a factory. Do the figures

$$14 \quad 9 \quad 11 \quad 13 \quad 8 \quad 12 \quad 13 \quad 16$$

suggest that the number of errors depends on the hour of the shift?

(4) An experiment involves tossing 4 coins 160 times and recording the number of heads on each toss. Test whether the coins can be considered as fair with the experimental figures tabulated.

Number of heads	Frequency
0	13
1	41
2	57
3	42
4	7

State clearly how you construct the test and any assumptions that you make.

(5) Complete Example 3 in the chapter about the fitting of a normal distribution to the absences distribution, using χ^2 to test the 'goodness of fit'.

(6) The table gives the results of an experiment with 80 people to test the effect of a particular vaccination in combating the disease Phlu. Test the hypothesis that the vaccination makes no difference to the catching of Phlu.

	Caught Phlu	Did not catch Phlu
Vaccinated	9	31
Not vaccinated	15	25

What is the effect on the test of using Yates' correction?

(7) The famous Market Research Agency Nobbut-Middling, Champion and Gradely investigates the Market share enjoyed by 3 leading brands of a particular product in the North, the Midlands and the South. The table gives the figures obtained from its samples. Test whether there is a significant difference in the brands' market share between the 3 regions considered.

	North	Midlands	South
Brand X	41	22	17
Brand Y	25	19	16
Brand Z	4	9	47

(8) The employees in 4 different firms are distributed in the 3 skill categories shown in the table. Use χ^2 to test the hypothesis that there is no connection between the firm and the type of labour.
Explain your reasoning clearly.

	Firm A	B	C	D
Skilled	24	24	23	49
Semi-skilled	32	60	37	51
Manual	24	56	40	80

22 Tests for Variance and the F Distribution

TESTS FOR VARIANCE

In earlier chapters we tested assorted hypotheses, including the important test on a mean and the test for the difference between two means. In the latter test we had to assume that unknown variances, of the two distributions from which the respective samples were taken, could be considered as equal. We are about to consider a way of testing two variances for equality with a new distribution called the *F distribution*, but first we shall discuss a test for one variance using our χ^2 distribution. We said in the last chapter that it had a link with variance, and the similar method of its construction hinted at this. By considering a simple example of the binomial distribution we can establish the link.

Suppose that n trials are carried out of an experiment which has probability p of a success on any individual trial. The probability of an individual failure is

$$q = 1 - p$$

If the number of successes that we *observe* in the n trials is S_1, then F_1 is the number of failures, where

$$S_1 + F_1 = n$$

The number of successes that we *expect*, however, is np; for example, an individual success probability of $\tfrac{3}{10}$ over 50 trials gives an expected number of successes of 15. Similarly, the expected number of failures is

$$nq = n(1-p)$$

These terms are summarised in Table 22.1.

Table 22.1

	Successes	Failures	Total
Observed, O	S_1	F_1	n
Expected, E	np	nq	n

The χ^2 criterion to measure the discrepancy therefore gives

$$\sum \frac{(O-E)^2}{E} = \frac{(S_1 - np)^2}{np} + \frac{(F_1 - nq)^2}{nq}$$

Substituting for F_1, we get

$$\sum \frac{(O-E)^2}{E} = \frac{(S_1 - np)^2}{np} + \frac{[n - S_1 - n(1-p)]^2}{nq}$$

$$= \frac{(S_1 - np)^2}{np} + \frac{(n - S_1 - n + np)^2}{nq}$$

$$= \frac{(S_1 - np)^2}{np} + \frac{(S_1 - np)^2}{nq}$$

since we can square out the minus factor in the second bracket. Hence,

$$\sum \frac{(O-E)^2}{E} = (S_1 - np)^2 \left(\frac{1}{np} + \frac{1}{nq} \right)$$

$$= (S_1 - np)^2 \left(\frac{q+p}{npq} \right)$$

$$= \frac{(S_1 - np)^2}{npq} \quad (\text{as } q + p = 1)$$

We shall now look closely at each entry in the right hand side (r.h.s.), remembering that we are dealing with a binomial distribution with probability p in a sample size n. In fact,

$S_1 = x$ (i.e. number of successes in the sample)

$np = \mu$ (i.e. mean number of successes)

$npq = \sigma^2$ (i.e. variance of the distribution)

(You can revise this from the binomial distribution section in Chapter 14.) Substituting, we get

$$\chi^2 = \sum \frac{(O-E)^2}{E} = \frac{(x - \mu)^2}{\sigma^2} \quad \left[= \left(\frac{x - \mu}{\sigma} \right)^2 \right]$$

There are several things to notice about this connection.

The χ^2 summation on the left hand side (l.h.s.) is made over two categories or terms, whereas the r.h.s. is in this case only over one term. Notice that the number of degrees of freedom on the l.h.s. is only 1, since

$$\nu = k - 1$$

because there is one restriction on total frequency and k is 2. The r.h.s. is therefore distributed as χ^2 with 1 degree of freedom.

Notice from the square bracket that it is really the square of a standardised normal variable $(x - \mu)/\sigma$, such as we met in discussing the normal distribution (see Chapter 15) and have used as the basis for most of our tests.

The general connection is that χ^2 is the sum of squared standardised normal deviations, the number of which is the number of degrees of freedom. They can each, in general, come from different populations; that is,

$$\chi^2 = \sum_{}^{n} \left(\frac{x - \mu}{\sigma}\right)^2 \quad \text{(with } n \text{ degrees of freedom)}$$

In the case where we take a sample of n objects from a normal distribution and measure a certain dimension x, say, with known mean μ, the sample variance s^2 will be given by

$$s^2 = \frac{\sum_{}^{n}(x - \mu)^2}{n} \quad \text{(as usual)}$$

$$\therefore ns^2 = \sum_{}^{n}(x - \mu)^2$$

$$\therefore \frac{ns^2}{\sigma^2} = \sum_{}^{n}\frac{(x - \mu)^2}{\sigma^2}$$

Thus, ns^2/σ^2 is distributed as χ^2 with n degrees of freedom.

We are saying that, if we take a sample size n from a normal population and calculate its sample variance, n times the ratio of the sample variance to the population variance is one value of a χ^2 distribution with n degrees of freedom. This enables us:

(1) to estimate σ for a population, using s and a confidence interval; and also

(2) to test for a change in a supposed value of σ, using s and some critical value of the χ^2 distribution.

Example 1

It is known from past experience that the mean diameter of a certain mechanical component is 30 mm. A random sample of 6 such components gives diameters (in millimetres) of

$$28 \quad 31 \quad 28 \quad 27 \quad 32 \quad 28$$

Assuming that the population diameters follow a normal distribution, find a 95% confidence interval for the population variance of diameters. Also, test the belief that the population variance is 3 mm².

From the calculations in Table 22.2 we find that $s^2 = \frac{26}{6}$;

$$\therefore \quad ns^2 = 26$$

$$\therefore \quad \frac{ns^2}{\sigma^2} = \frac{26}{\sigma^2}$$

Table 22.2 *Calculation of sample variance ($\mu = 30$, $n = 6$)*

x	28	31	28	27	32	28
$x - \mu$	-2	1	-2	-3	2	-2
$(x - \mu)^2$	4	1	4	9	4	4

$$s^2 = \frac{\Sigma(x-\mu)^2}{n}$$

$$= \tfrac{26}{6} \quad (= 4.5)$$

Note: $\bar{x} = 29$.

If we take a central confidence interval for ns^2/σ^2 in the χ^2 distribution, we get two $2\tfrac{1}{2}$% tails at a and b as shown in Figure 22.1. From Table 8, we find that

$$a = 1.237 \quad b = 14.449$$

Therefore, we are 95% confident that a χ^2 value will lie between a and b; that is, we are 95% confident that

$$a < \frac{ns^2}{\sigma^2} < b \quad \text{since} \quad \frac{ns^2}{\sigma^2} \sim \chi^2$$

$$\therefore \; 1.237 < \frac{26}{\sigma^2} < 14.449$$

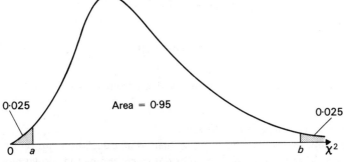

Figure 22.1 χ^2 distribution ($\nu = 6$)

Cross-multiplying gives

$$\sigma^2 < \frac{26}{1.237} \quad \text{and} \quad \frac{26}{14.449} < \sigma^2$$

$$\therefore \frac{26}{14.449} < \sigma^2 < \frac{26}{1.237}$$

$$1.799 < \sigma^2 < 21.019$$

That is, a 95% CI for σ^2 is 1.799 to 21.019. (*Note*: If it is the standard deviation σ that is required to have the CI, just take the square root of each figure.)

To test the belief that $\sigma^2 = 3$ we set up hypotheses thus:

$$H_0: \sigma^2 = 3 \quad \text{(i.e. population variance is 3)}$$

$$H_A: \sigma^2 > 3 \quad \text{(i.e. a one-tail test)}$$

Thus,

$$\frac{ns^2}{\sigma^2} = \frac{26}{3} = 8.67$$

At the 0.05 level (one-tail test) with 6 degrees of freedom, Table 8 of Murdoch and Barnes* gives a χ^2 value of 12.592. Our value is within this critical value, so we accept H_0. That is, the evidence suggests that the belief that $\sigma^2 = 3$ is justified.

Notice that $\nu = 6$ in Example 1 because we knew μ and therefore had no restrictions. In an example where μ is not known and you have to calculate s^2, it will be necessary to use

*See Neave, page 21.

$$s^2 = \frac{\Sigma(x - \bar{x})^2}{n}$$

instead of

$$s^2 = \frac{\Sigma(x - \mu)^2}{n}$$

The restriction also necessitates subtracting 1 from n, to give

$$\nu = n - 1$$

Otherwise, the procedure is the same.

In many examples s^2 is given, so there is no calculation; *but* remember to check whether you are given μ or not when finding the correct number of degrees of freedom (see Exercise 1).

THE F DISTRIBUTION

We use the F distribution to test a claim about the population variances of two normal populations. In particular, we can test for the equality of two variances coming from different populations, something that we formerly had to assume when testing the difference of two means.

We define the F distribution as follows:

If
$$x \sim N(\mu_x, \sigma_x^2) \quad \text{and} \quad y \sim N(\mu_y, \sigma_y^2)$$
$$x \sim N(\mu_x, \sigma_x^2) \quad \text{and} \quad y \sim N(\mu_y, \sigma_y^2)$$

and samples size n and N respectively are taken, which have sample variances s_x^2 and s_y^2, then the ratio

$$\frac{s_x^2}{\sigma_x^2} \bigg/ \frac{s_y^2}{\sigma_y^2} \sim F_{n,N} \qquad \left(= \frac{s_x^2}{\sigma_x^2} \cdot \frac{\sigma_y^2}{s_y^2} \right)$$

A ratio is the obvious measure, because it is the relative magnitude of variances that interests us.

When σ_x^2 and σ_y^2 are assumed equal (e.g. under a null hypothesis), the ratio reduces to

$$\frac{s_x^2}{s_y^2} \sim F_{n,N}$$

where n and N are the degrees of freedom due to the variance from each original distribution. These are referred to in Table 9 of Murdoch and Barnes[*] as ν_1 and ν_2 respectively; and if μ_x and μ_y are known, they are equal to n and N.

[*]See Neave, page 22.

If μ_x and μ_y are unknown (the usual case), we adapt just as we did for the single variance test earlier in this chapter, with one addition. The calculation (if necessary) is made of the unbiased estimates of variance $\hat{\sigma}_x^2$ and $\hat{\sigma}_y^2$, using $\bar{x}$ and $\bar{y}$ respectively instead of μ_x and μ_y, and the degrees of freedom become

$$\nu_1 = n - 1 \quad \text{and} \quad \nu_2 = N - 1$$

due to our restrictions.

A confidence interval for σ_y^2/σ_x^2 can be found with the F distribution, just as for a single σ^2 above (with the χ^2 distribution), by using the fact that the ratio

$$\frac{\hat{\sigma}_x^2}{\hat{\sigma}_y^2} \left(\frac{\sigma_y^2}{\sigma_x^2} \right)$$

will lie a certain percentage of the time between two marks a and b of the F distribution.

The shape of the F distribution (see Figure 22.2) varies for different values of ν_1 and ν_2, but you should see that it will be similar to the χ^2 shape, because the ratios $\hat{\sigma}_x^2/\hat{\sigma}_y^2$ that are possible (in a particular case of n and N, with $\sigma_x = \sigma_y$) vary between zero and theoretically, any value.

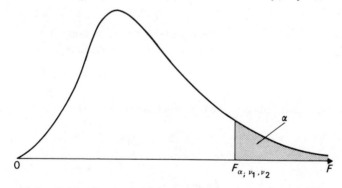

Figure 22.2 F distribution with degrees of freedom ν_1 (in numerator) and ν_2 (in denominator)

One convenience in the tabulation of F that you should note is that the larger sample variance is always put in the numerator, so the ratio is always greater than 1. The larger variance is then the one with ν_1 degrees of freedom.

Example 2

Two laboratory assistants are tested in the use of identical new machines to measure the concentration of a certain chemical. Random samples of their work show respectively a standard deviation of 3 units in a sample of 10 measurements and a standard deviation of 4 units in a sample of 6 measurements. Do these results indicate that one assistant is more accurate than the other?

We are not given μ_x and μ_y, and we must assume that the sample deviations given are s_x and s_y, not $\hat{\sigma}_x$ and $\hat{\sigma}_y$. Therefore, we have

$$n = 10 \qquad N = 6$$
$$s_x^2 = 9 \qquad s_y^2 = 16$$

and we use

$$\hat{\sigma}_x^2 = \frac{ns_x^2}{n-1} \qquad \hat{\sigma}_y^2 = \frac{ns_y^2}{n-1}$$

(correcting as usual to get the unbiased estimators). Substituting gives

$$\hat{\sigma}_x^2 = \tfrac{10}{9} \times 9 \qquad \hat{\sigma}_y^2 = \tfrac{6}{5} \times 16$$
$$= 10 \qquad\qquad = 19.2$$
$$\therefore \nu_2 = 9 \qquad \nu_1 = 5$$

making the larger variance the numerator (and therefore having ν_1). The test is

H_0: $\qquad \sigma_x = \sigma_y \qquad$ (i.e. equally accurate, meaning that measurements are from same population)

H_A: $\qquad \sigma_x < \sigma_y \qquad$ (i.e. second assistant less accurate)

Using the F distribution at the 0.05 level, we compare our ratio

$$\frac{\hat{\sigma}_y^2}{\hat{\sigma}_x^2} = \frac{19.2}{10} = 1.92$$

with

$$F_{0.05;\, 5,\, 9} = 3.48$$

The value 1.92 is well within 3.48 and therefore could easily have come about by chance, due to random sampling errors, so we accept H_0. We conclude that the accuracy of the assistants is not significantly different.

We shall add to Example 2 by demonstrating how to calculate the lower points of the F distribution. This is also described in Table 9. For example, to find the value having the 5% lower tail below it we need $F_{0.95;\, 5,\, 9}$ and use the relationship

Tests for Variance and the F Distribution

$$F_{1-\alpha; \nu_1, \nu_2} = \frac{1}{F_{\alpha; \nu_2, \nu_1}}$$

$$\therefore F_{0.95; 5, 9} = \frac{1}{F_{0.05; 9, 5}}$$

$$= \frac{1}{4.78}$$

$$= 0.21$$

Notice that it was necessary to interpolate for $\nu_1 = 9$.

One other feature of the F table is the way that ν_1 and ν_2 are given up to ∞ (infinity). This enables us to test such a situation as Example 1 with the F distribution, as if the sample had yielded

$$\hat{\sigma}_x^2 = \sum \frac{(x - \bar{x})^2}{n - 1} = \frac{20}{5}$$

with 5 degrees of freedom (check it!), while a very large sample with infinite degrees of freedom had given an estimated variance of 3. We are assuming now that μ was not known, so $\bar{x} = 29$ is used to calculate $\hat{\sigma}_x^2$. Thus

$$\hat{\sigma}_x^2 = 4 \qquad \hat{\sigma}_y^2 = 3$$

$$\nu_1 = 5 \qquad \nu_2 = \infty$$

$$\therefore \frac{\hat{\sigma}_x^2}{\hat{\sigma}_y^2} = \frac{4}{3} = 1.33$$

From Table 9,

$$F_{0.05; 5, \infty} = 2.21$$

so again we have a non-significant result.

We shall conclude by testing our previous Example 3 of Chapter 20 to check our assumption that the unknown variances were equal. The same test should be done in Exercises 3 and 4 before concluding that the difference-of-means test can be carried out, although, if it *is* justified, you should then go on to do so.

Example 3
Table 22.3 gives the relevant figures for the two samples. We calculated (see Table 20.2) that

Table 22.3

x	46	45	48	44	47	44	48	46	
y	43	44	46	45	45	46	44	46	44	47

$$\Sigma(x-\bar{x})^2 = 18 \qquad \Sigma(y-\bar{y})^2 = 14$$

where $n = 8$ and $N = 10$. Thus

$$\hat{\sigma}_x^2 = \tfrac{18}{7} \qquad \hat{\sigma}_y^2 = \tfrac{14}{9}$$

$$\nu_1 = 7 \qquad \nu_2 = 9$$

We want to test

$$H_0: \quad \sigma_x = \sigma_y \quad (= \sigma, \text{i.e. unknown but equal})$$
$$H_A: \quad \sigma_x > \sigma_y \quad (\text{i.e. a one-tail test})$$

Our ratio is

$$\frac{\hat{\sigma}_x^2}{\hat{\sigma}_y^2} = \frac{18}{7} \Big/ \frac{14}{9} = \frac{162}{98} = 1.653$$

Using the F distribution at the 0.05 level, we compare this with (from Table 9)

$$F_{0.05;\,7,\,9} = 3.29$$

our result is not significant, so our assumption of equal variances is justified. We can therefore proceed with our t test on the difference between the means $\bar{x}$ and $\bar{y}$. Remember that it was still necessary to estimate the unknown σ by $\hat{\sigma}$, using the estimates from the samples.

EXERCISES

(1) A random sample of 30 students from a large college gives a standard deviation of IQ of 12 points. Calculate 95% confidence limits for the standard deviation of IQ of all the students in the college.

(2) In the Cruncho cereal factory a filling machine is modified to reduce the variability of packet contents. From long experience it is known that the machine fills to a mean weight of 680 g with a standard deviation of 6.2 g. After modification (which is assumed not to alter the mean weight) a random sample of 20 packets is found to have a standard deviation of 4.3 g. Is this decrease in variability significant at the 0.05 level? (*Hint:* Which tail are you investigating?)

(3) The marks of a sample of 17 male students in a statistics examination show a mean of 51 with a standard deviation (unbiased) of 5.2. A

sample of 25 female students gives a mean of 60 with a standard deviation (unbiased) of 9.8. Test the hypothesis that the variation of marks is the same for male or female students; and if it is valid to do so, test for a difference between the mean marks.

(4) A random sample is taken of the wages of 9 men in a particular industry in Scotland, with the following results (in £/week):

    68   69   77   81   74   76   79   76   75

Another random sample of 10 men in the same industry in England gives the figures

    73   85   72   89   86   88   77   73   82   85

Do these figures suggest that the sample of Englishmen has a greater variation in the wages?

Is there evidence of a difference in the average weekly wage between England and Scotland?

State clearly any assumptions that you make.

23 Analysis of Variance

ONE-WAY ANALYSIS OF VARIANCE

We have previously tested to see whether a difference between two means was significant or not. This was done with a null hypothesis that the means were equal and an alternative hypothesis that they were not (i.e. a two-tail test); only a test statistic to measure the difference that had a significantly large value was considered to be capable of rejecting the null hypothesis. Otherwise, any difference was taken as due to chance (i.e. random experimental differences between the figures).

We can now extend this idea to test whether observed differences in more than two means are caused by real differences or purely by chance. Consider the following situation.

Example 1

A certain type of 1,300 cc car is used to make 4 test drives with exactly 1 gallon of each of 3 different petrol formulations A, B and C. The 12 mileages x shown in Table 23.1 are obtained.

Table 23.1 Mileage

A	B	C
31	27	34
34	29	31
32	30	35
35	30	36

Note: $k = 3$ columns and $n = 4$ rows (i.e. the sample size for each petrol).

Are the differences in mean m.p.g. significant or attributable to chance?

We shall assume for the moment that there is no other factor (e.g. a difference in the cars used) that could have caused the variation in m.p.g.; only the difference in petrol or random errors could have done so.

Since we are basically testing the overall variability of the figures, what we need is two independent estimates of the overall variance, which can be checked one against the other. The F distribution that we studied in Chapter 22 provides a way of comparing two variances, but how can we get two estimates from the one set of figures?

Well, each petrol formulation has its own mean m.p.g. and its own variance 'within the column' (i.e. of its own sample of figures). We can calculate these, as shown in Table 23.2. It is important that these variances are not too dissimilar, because we need to assume that the samples are from populations with the same variance σ^2. Thus, for each petrol we have a mean and a variance.

Let us attempt to estimate the overall variance by finding the mean of the three variances and the variance of the three means. Notice that we use $k = 3$ (columns) rather than $n = 4$, because there are three petrol types.

The (unbiased) *variance of the means*, which have grand mean $\bar{\bar{x}} = 32$ found by

$$\bar{\bar{x}} = \frac{\bar{x}_A + \bar{x}_B + \bar{x}_C}{k}$$

is given by

$$\sigma_{\bar{x}}^2 = \frac{\overset{k \text{ means}}{\underset{\text{of A, B, C}}{\Sigma}} (\bar{x} - \bar{\bar{x}})^2}{k - 1}$$

$$= \frac{(33 - 32)^2 + (29 - 32)^2 + (34 - 32)^2}{3 - 1}$$

$$= \frac{1 + 9 + 4}{2}$$

$$= 7$$

Now, we know from our work on the standard error that the variance $\sigma_{\bar{x}}^2$ of a sample means distribution is linked to the variance σ^2 of the original distribution from which the samples size n are taken by

$$\sigma_{\bar{x}}^2 = \frac{\sigma^2}{n}$$

Since we have $\hat{\sigma}_{\bar{x}}^2 = 7$ and $n = 4$,

$$\hat{\sigma}_{\text{from means}}^2 = 4 \times 7 = 28$$

is our first estimate of the overall variance.

The *mean of the variances* is simply

$$\frac{\hat{\sigma}_A^2 + \hat{\sigma}_B^2 + \hat{\sigma}_C^2}{k} = \frac{3.33 + 2 + 4.67}{3} = \frac{10}{3} = 3.33$$

Table 23.2 Calculation of mean and variance (unbiased estimate) for each petrol type ($n = 4$)

x_A	$x_A - \bar{x}_A$	$(x_A - \bar{x}_A)^2$	x_B	$x_B - \bar{x}_B$	$(x_B - \bar{x}_B)^2$	x_C	$x_C - \bar{x}_C$	$(x_C - \bar{x}_C)^2$
31	−2	4	27	−2	4	34	0	0
34	1	1	29	0	0	31	−3	9
32	−1	1	30	1	1	35	1	1
35	2	4	30	1	1	36	2	4
132		10	116		6	136		14

$$\bar{x} = \frac{\Sigma x}{n}$$

$$\therefore \bar{x}_A = \frac{132}{4} = 33$$

$$\bar{x}_B = \frac{116}{4} = 29$$

$$\bar{x}_C = \frac{136}{4} = 34$$

$$\hat{\sigma}^2 = \frac{\sum_{n}(x - \bar{x})^2}{n - 1}$$

$$\therefore \hat{\sigma}_A^2 = \tfrac{10}{3} = 3.33$$

$$\hat{\sigma}_B^2 = \tfrac{6}{3} = 2$$

$$\hat{\sigma}_C^2 = \tfrac{14}{3} = 4.67$$

That is, we have another estimate

$$\hat{\sigma}^2_{\text{from variances}} = 3.33$$

of the overall variance.

We can now use F to compare them with the variance ratio $28/3.33$, which is just

$$\frac{\text{Estimate of } \sigma^2 \text{ from variation between the (sample) means}}{\text{Estimate of } \sigma^2 \text{ from variation within the samples}}$$

The relevant degrees of freedom are

Numerator: $\quad \nu_1 = k - 1$
$\quad\quad\quad\quad\quad\quad\quad = 3 - 1 = 2$
Denominator: $\quad \nu_2 = k(n-1)$
$\quad\quad\quad\quad\quad\quad\quad = 3(4-1) = 9$

$\nu_1 = 2$ because we calculated our first estimate from three (k) means with one restriction on the unknown true grand mean; and $\nu_2 = 9$ because the second estimate was calculated from the average of three (k) variances, *each of which* came from a sample that had four (n) figures and one restriction of an unknown mean.

Thus, we get

$$F_{2,9} = \frac{28}{3.33} = 8.4$$

From Table 9 of Murdoch and Barnes we find a suitable critical value

$$F_{0.01;2,9} = 8.02$$

Our value exceeds this, so it seems that there is a significant difference between the petrols at the 0.01 level.

The variation *between* each petrol (formally called the treatments factor) is significant compared to the variation *within* the petrol samples (formally called the error factor). The use of the word 'treatment' is due to early work in agricultural studies using analysis of variance to test the effect of, for example, different fertilisers.

Once we have decided that the petrols are not all the same, it is reasonable to wonder if we can identify which of the differences in means in our example is significant.

We can use the t test as before to test each pair, from

$$\bar{x}_A = 33 \quad\quad \bar{x}_B = 29 \quad\quad \bar{x}_C = 34$$

but we need an estimate $\hat{\sigma}^2$ of the overall variance σ^2 in order to use our statistic

$$t = \frac{\bar{x} - \bar{y}}{\sqrt{\left[\hat{\sigma}^2\left(\frac{1}{n} + \frac{1}{N}\right)\right]}}$$

Remember that under

H_0 : $\mu_x = \mu_y$ (i.e. true means equal)

H_A : $\mu_x \neq \mu_y$ (i.e. unequal, $\therefore$ two-tail test)

The best estimate that we have is that from the 'within samples' variation, namely,

$$\hat{\sigma}^2_{\text{from variances}} = 3.33$$

because there is no 'between samples' difference present.

Note very carefully that the degrees of freedom are $\nu = 9$ for this variance, because of its method of derivation.

Substituting gives

$$t = \frac{\bar{x} - \bar{y}}{\sqrt{(\frac{10}{3} \times \frac{1}{2})}}$$

since $n = N = 4$ for each sample.

We could test each pair of means in turn, but it is simpler to select a suitable t value and see what difference is needed to be critical. Suppose that we take, from Table 7 of Murdoch and Barnes,

$$t_{0.005;9} = 3.25$$

Substituting and rearranging, we get

$$3.25\sqrt{(\tfrac{10}{3} \times \tfrac{1}{2})} = \bar{x} - \bar{y} \quad \text{(critical at 0.01 level)}$$

which yields

$$\text{Critical } (\bar{x} - \bar{y}) = 3.25 \times 1.291 = 4.2$$

We had differences of $4(= \bar{x}_A - \bar{x}_B)$, $5(= \bar{x}_C - \bar{x}_B)$ and $1(= \bar{x}_C - \bar{x}_A)$, so the difference between the mean m.p.g. of 29 for petrol B and 34 for petrol C is significant at the 0.01 level.

Notice that the level of significance that we have chosen is high for both the F test and the t test; a different choice could possibly give a different result.

Analysis of Variance

We can now put the work of Example 1 on a formal footing, known as one-way analysis of variance.

We are attempting to express the total variability of a set of figures, such as we had for m.p.g., as the *sum of terms* that are separately attributable to specific factors or causes of the variation. We considered the two factors of actual differences in petrol and errors due to chance. Remember that even the same car with the same petrol used in the same conditions will always have some chance or random variation over different runs. It is a basic assumption that these random errors on every figure are normally distributed.

Later, we shall even extend the analysis to a further factor possibly producing variation, but for now we handle the figures using the **sum of squares** SS for the factors of treatment, error and total. In the usual notation these are SS_{Tr}, SS_E and SS_T respectively.

Any variance is of course closely linked to a sum of squares (of deviations), and what we actually had for treatments was

$$\hat{\sigma}^2_{\text{from means}} = n\hat{\sigma}^2_{\bar{x}} = \frac{n \overset{k \text{ means of A, B, C}}{\Sigma} (\bar{x} - \bar{\bar{x}})^2}{k-1} = \frac{SS_{Tr}}{k-1}$$

The sum of squared deviations here is for those between each mean and the grand mean $\bar{\bar{x}}$. Remember that

$$\bar{\bar{x}} = \frac{\overset{A, B, C}{\Sigma} \bar{x}}{3} = \frac{\overset{\text{all } x}{\Sigma} x}{12} \quad \left(= \frac{G}{nk} \quad \text{(see later)} \right)$$

For the errors (within) we had

$$\hat{\sigma}^2_{\text{from variances}} = \frac{1}{k} \left(\frac{\overset{n}{\Sigma}(x_A - \bar{x}_A)^2}{n-1} + \frac{\overset{n}{\Sigma}(x_B - \bar{x}_B)^2}{n-1} + \frac{\overset{n}{\Sigma}(\bar{x}_C - x_C)^2}{n-1} \right)$$

$$= \frac{\overset{k \text{ petrols}}{\Sigma} [\overset{n}{\Sigma}(x - \bar{x})^2]}{k(n-1)}$$

$$= \frac{SS_E}{k(n-1)}$$

The total variance is

$$\hat{\sigma}^2 = \frac{\overset{\text{all } x}{\Sigma}(x - \bar{\bar{x}})^2}{nk-1} = \frac{SS_T}{nk-1}$$

The basic connection is that

$$SS_T = SS_{Tr} + SS_E$$

which neatly expresses the fact that the total sum of squares can be expressed as the sum of two separate factors. It is true because, substituting for each SS from the above, we get

$$\sum_{\text{all } x} (x - \bar{\bar{x}})^2 = n \sum_{\substack{k \text{ means} \\ \text{of A, B, C}}} (\bar{x} - \bar{\bar{x}})^2 + \sum_{\substack{k \\ \text{petrols}}} \left[\sum_{n} (x - \bar{x})^2 \right]$$

which is a mathematical identity. Even stating it rigorously with full subscript notation is frightening. Consequently, we shall leave a proof until the end of this chapter, but we can verify its truth for our example.

We have of course worked out the two terms of the r.h.s. already, namely,

$$\left. \begin{array}{l} SS_{Tr} = n \sum_{\substack{k \text{ means} \\ \text{of A, B, C}}} (\bar{x} - \bar{\bar{x}})^2 \\ \\ \quad\quad = 4(1 + 9 + 4) = 56 \\ \\ SS_E = \sum_{\substack{k \\ \text{petrols}}} \left[\sum_{N} (x - \bar{x})^2 \right] \\ \\ \quad\quad = 10 + 6 + 14 = 30 \end{array} \right\} \text{Sum 86}$$

To find SS_T we must first list the deviation of every figure from the grand mean $\bar{\bar{x}}$. (Check that $\bar{\bar{x}} = 32$ by adding every figure and dividing by 12.) The deviations of every original figure from 32 are

$$\left. \begin{array}{rrr} -1 & -5 & 2 \\ 2 & -3 & -1 \\ 0 & -2 & 3 \\ 3 & -2 & 4 \end{array} \right\} \text{ which square to } \left\{ \begin{array}{rrr} 1 & 25 & 4 \\ 4 & 9 & 1 \\ 0 & 4 & 9 \\ 9 & 4 & 16 \end{array} \right.$$

Total = 86

$$\therefore SS_T = \sum_{\text{all } x} (x - \bar{\bar{x}})^2 = 86$$

Each formula for the SS can be expanded for ease of calculation, just as our formula

$$\frac{\Sigma(x-\bar{x})^2}{n}$$

converted to the equivalent formula

$$\text{Variance} = \frac{\Sigma x^2}{n} - \bar{x}^2$$

(see the proof at the end of the chapter). We get the converted formulae

$$SS_T = \sum_{}^{\text{all } x} x^2 - nk\bar{\bar{x}}^2 \qquad (nk = \text{total number of figures})$$

or

$$SS_T = \sum_{}^{\text{all } x} x^2 - \frac{G^2}{nk} \qquad (G = \text{sum of all figures})$$

and

$$SS_{Tr} = \frac{\sum_{}^{\substack{k \\ \text{treatments}}} C_t^2}{n} - \frac{G^2}{nk} \qquad (C_t = \text{subtotal for each column, i.e. petrol or treatment})$$

Hence, by subtraction we easily find

$$SS_E = SS_T - SS_{Tr}$$

This notation enables us to summarise everything that we did in Example 1 with what is called the **analysis of variance (ANOVA) table** (see Table 23.3).

Table 23.3 *One-way ANOVA table*

Variation factor	Degrees of freedom, ν	Sum of squares	$\frac{SS}{\nu}$ = Variance	F value
Treatment	$k-1$	SS_{Tr}	$\frac{SS_{Tr}}{k-1} = V_{Tr}$	$\frac{V_{Tr}}{V_E}$
Error	$k(n-1)$	SS_E	$\frac{SS_E}{k(n-1)} = V_E$	
Total	$nk-1$	SS_T		

We shall demonstrate the method with this table by working through our Example 1 again as economically as possible and, of course, using the converted formulae. This is how you should always do an exercise, but it is important to understand the completely equivalent but more basic working that we first used to solve the example. We shall also demonstrate that subtracting a suitable large number (e.g. 27) from every (m.p.g.) figure does not affect the solution. This is because a variance is a measure of *relative* deviation and is unaffected by location. The new example, although really unchanged, is given in Table 23.4.

Table 23.4 *Calculation of terms for SS formulae ($n = 4$ rows, $k = 3$ columns)*

Mileage, x (+ 27)			Squares		
A	B	C	A	B	C
4	0	7	16	0	49
7	2	4	49	4	16
5	3	8	25	9	64
8	3	9	64	9	81
24	8	28	Total = 386		

(1) $C_A = 24$, $C_B = 8$, $C_C = 28$

$$\therefore G = \sum_{\text{treatments}}^{k} C_t = 60$$

(2) $C_A^2 = 576$, $C_B^2 = 64$, $C_C^2 = 784$

$$\therefore \sum_{\text{treatments}}^{k} C_t^2 = 1{,}424$$

(3) $\sum^{\text{all } x} x^2 = 386$

(4) $nk = 12$

Hence, we get

$$SS_T = \sum^{\text{all } x} x^2 - \frac{G^2}{nk}$$

$$= 386 - \frac{3{,}600}{12} = 86 \quad \text{(as before)}$$

$$SS_{Tr} = \frac{\overset{k}{\underset{}{\Sigma}}\, C_t^2}{n} - \frac{G^2}{nk}$$

$$= \frac{1{,}424}{4} - 300 = 56 \quad \text{(as before)}$$

$$\therefore\ SS_E = SS_T - SS_{Tr}$$

$$= 86 - 56 = 30 \quad \text{(as before)}$$

We have

H_0: No difference in petrols (means)

H_A: There is a difference

Substituting in the ANOVA table (see Table 23.5), we get $F = 8.4$. We compare at the 0.01 level our F value of 8.4 with

$$F_{0.01;2,9} = 8.02$$

from Table 9 (see Figure 23.1). As before, we get 8.4 which exceeds 8.02, so we reject H_0. It seems that there is a difference in mean m.p.g. between the petrols.

Table 23.5

Variation factor	Degrees of Freedom, ν	Sum of Squares	$\frac{SS}{\nu}$ = Variance	F value
Treatment	2	56	$\frac{56}{2} = 28$	$\frac{28}{3.33} = 8.4$
Error	9	30	$\frac{30}{9} = 3.33$	
Total	11	86		

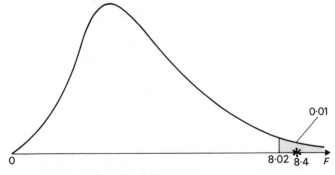

Figure 23.1 F distribution ($\nu_1 = 2, \nu_2 = 9$)

TWO-WAY ANALYSIS OF VARIANCE

It may have occurred to you from looking at what we did to the columns that something analogous may be appropriate for the rows. This is in fact the case when we have genuine reason to believe that some other factor is at work causing variation but is disguised in SS_E. Then we use a **two-way analysis of variance**.

Suppose in our Example 1 that the m.p.g. figures came not from one car or a random allocation of cars but from 4 particular cars, each test driven with a gallon of A, B and C. The new example (with 27 subtracted from each mileage figure) is given in Table 23.6.

Table 23.6 *Mileage (+ 27)*

		Petrol A	B	C
	I	4	0	7
Car	II	7	2	4
	III	5	3	8
	IV	8	3	9

Remember that it is very important to apply the one-way or two-way analysis-of-variance procedures from what you know collaterally about the design of your experiment; read an exercise very carefully. Only if you know that there is a real second factor (known, for agricultural reasons again, as **blocks**) possibly affecting the variation in the rows should you apply the two-way ANOVA.

What we really do is to test for differences in the means of the rows in a way that is analogous to our previous test on the columns. More elegantly, our previous result is extended to

$$SS_T = SS_{Tr} + SS_B + SS_e$$

where SS_B is the sum of squares due to blocks. The total SS is now expressed as the sum of squares due to three possible factors of variation: treatments (i.e. petrols or between columns), blocks (i.e. cars or between rows) and errors.

The formula for SS_B is completely analogous to that for SS_{Tr}, namely,

$$SS_B = \frac{\sum_{}^{n\ \text{Blocks}} R_b^2}{k} - \frac{G^2}{nk}$$

where R_b is the subtotal for each row (i.e. car or block).

The previous error factor is thus altered, since we have effectively extracted another factor from it, but the new SS for error is easily found from

$$SS_e = SS_T - SS_{Tr} - SS_B$$

We shall finally work through the whole example to show this, making the extra assumption of course about the 4 cars in the experiment. Using the figures in Table 23.7, we get

$$SS_T = \sum^{\text{all } x} x^2 - \frac{G^2}{nk}$$

$$= 386 - \frac{3{,}600}{12} = 86$$

$$SS_{Tr} = \frac{\sum\limits^{\text{treatments}}_{} C_t^2}{n} - \frac{G^2}{nk}$$

$$= \frac{1{,}424}{4} - 300 = 56$$

$$SS_B = \frac{\sum\limits^{\text{blocks}}_{} R_b^2}{k} - \frac{G^2}{nk}$$

$$= \tfrac{946}{3} - 300 = 15.33$$

$$\therefore SS_e = SS_T - SS_{Tr} - SS_b$$

$$= 86 - 56 - 15.33 = 14.67$$

We have

H_0: No difference in petrols or cars ($\therefore$ really two tests, see below)

H_A: A difference in petrols (test one) and/or a difference in cars (test two)

The two-way ANOVA table and relevant substitutions are given in Tables 23.8 and 23.9 respectively. We take $\alpha = 0.01$. Hence,

Test one: Compare 11.45 with

$$F_{0.01;2,6} = 10.92$$

∴ Significant

Test two: Compare 2.09 with

$$F_{0.01;3,6} = 9.78$$

∴ Not significant

It seems that with our experimental design assumptions about the use of petrols and cars that there is a real difference due to petrols but not a real difference due to cars.

Table 23.7 *Calculation of terms for SS formulae ($n = 4$ rows, $k = 3$ columns)*

	Mileage, x (+ 27)				Squares		
	A	B	C		A	B	C
I	4	0	7	$R_I = 11$	16	0	49
II	7	2	4	$R_{II} = 13$	49	4	16
III	5	3	8	$R_{III} = 16$	25	9	64
IV	8	3	9	$R_{IV} = 20$	64	9	81
	24	8	28		Total = 386		

(1) $C_A = 24$, $C_B = 8$, $C_C = 28$

$$\therefore G = \sum_{t}^{\substack{k \\ \text{treatments}}} C_t = 60$$

(2) $C_A^2 = 576$, $C_B^2 = 64$, $C_C^2 = 784$

$$\therefore \sum^{\substack{k \\ \text{treatments}}} C_t^2 = 1{,}424$$

(3) $R_I^2 = 121 \quad R_{II}^2 = 169 \quad R_{III}^2 = 256$

$R_{IV}^2 = 400$

$$\therefore \sum_{b}^{\substack{n \\ \text{blocks}}} R_b^2 = 946$$

(4) $\sum^{\text{all } x} x^2 = 386$

(5) $nk = 12$

Table 23.8 Two-way ANOVA table

Variation factor	Degrees of freedom, v	Sum of squares	$\dfrac{SS}{v}$ = Variance	F value
Treatment	$k-1$	SS_{Tr}	$\dfrac{SS_{Tr}}{k-1} = V_{Tr}$	$\dfrac{V_{Tr}}{V_\epsilon}$ (test one)
Block	$n-1$	SS_B	$\dfrac{SS_B}{n-1} = V_B$	$\dfrac{V_B}{V_\epsilon}$ (test two)
Error	$(n-1)(k-1)$	SS_ϵ	$\dfrac{SS_\epsilon}{(n-1)(k-1)} = V_\epsilon$	
Total	$nk-1$	SS_T		

Table 23.9

Variation factor	Degrees of freedom, v	Sum of squares	$\dfrac{SS}{v}$ = Variance	F value
Treatment	2	56	$\dfrac{56}{2} = 28$	$\dfrac{28}{2.445} = 11.45$ (test one)
Block	3	15.33	$\dfrac{15.33}{3} = 5.11$	$\dfrac{5.11}{2.445} = 2.09$ (test two)
Error	6	14.67	$\dfrac{14.67}{6} = 2.445$	
Total	11	86		

Remember that you must be very clear about the assumptions in your experimental design in order to pick the right test.

As a footnote, notice from the last ANOVA table (Table 23.8) that we could have done the F test using total variance V_T (i.e. using SS_T). However, it contains SS_{Tr} and SS_B and their possibly non-random variations. Consequently, V_ϵ is used, because SS_ϵ is independent of them.

DESIGN OF EXPERIMENTS

It became a legal requirement in 1978 that every new car must display in the showroom its miles-per-gallon figure under different conditions: in 'typical' traffic, at a steady 50 m.p.h., and so on. How are such m.p.g. figures arrived at? There is always variability due to factors that we suspect and even to some that we do not.

When a magazine like *Which?* tests several comparable washing-machines, it considers such variables as different powders, different washing temperatures, whether the water is hard or soft, different lengths of wash, and so on, in trying to decide which machine performs as the 'best'. It is easy to say what a result 'is'; but only if the other factors are controlled can the 'why', or cause of a result, be measured against one factor. The constraint on that conclusion of 'why' is that it will then only be valid under the prescribed conditions. It is possible to overcontrol a situation, giving limited conclusions.

What is needed is a design or scheme for the experiment where more than one factor is allowed to vary. Let us use our example of the test on three petrols A, B and C using 4 (similar) cars I, II, III and IV, with the slight modification that a m.p.g. figure is needed at two speeds: fast (F) and slow (S). Then

$$3 \times 4 \times 2 = 24$$

combinations of these factors are possible for any single test drive, and a balanced design for the experiment would be the 24 test drives listed in Table 23.10; for example, FIA means that car I uses petrol A at fast speed.

Table 23.10

F I A	F III A	S I A	S III A
F I B	F III B	S I B	S III B
F I C	F III C	S I C	S III C
F II A	F IV A	S II A	S IV A
F II B	F IV B	S II B	S IV B
F II C	F IV C	S II C	S IV C

Although this design covers all the possible combinations of test drive, it has been constructed systematically, as you have probably noticed. If the test drives were made in systematic order (e.g. all the F combinations first), an unsuspected factor (e.g. changing weather conditions) could cause a bias in the m.p.g. figures. The design therefore needs to be randomised, and a simple way would be to give every combination a number from 1 to 24 and draw from a hat to determine sequence. If time and money allowed, the whole scheme could be run through again, with a different random sequence.

Thus, we attempt to obtain figures from which a detailed analysis can be made of the factors involved in getting those figures. Example 1 demonstrated this with two factors. Although it is more complicated with three or more factors, a similar analysis can be made. Remember that the assumptions made about the populations from which the figures come (e.g. that the figures are normally distributed) should always be kept in mind.

CONVERSION OF ANOVA FORMULAE

Total

$$SS_T = \sum_{}^{\text{all } x} (x - \bar{\bar{x}})^2$$

$$= \sum_{}^{\text{all } x} (x^2 - 2x\bar{\bar{x}} + \bar{\bar{x}}^2)$$

$$= \sum_{}^{\text{all } x} x^2 - 2\bar{\bar{x}} \sum_{}^{\text{all } x} x + \sum_{}^{\text{all } x} \bar{\bar{x}}^2$$

$$= \sum_{}^{\text{all } x} x^2 - 2\bar{\bar{x}}G + nk\bar{\bar{x}}^2 \qquad \left(\text{since } G = \sum_{}^{\text{all } x} x, \text{ i.e. the sum of all the figures, of which there are } nk\right)$$

$$= \sum_{}^{\text{all } x} x^2 - 2 \cdot \frac{G}{nk} \cdot G + nk \cdot \frac{G^2}{n^2 k^2} \qquad \left(\text{since } \bar{\bar{x}} = \frac{\Sigma x}{nk} = \frac{G}{nk}\right)$$

$$= \sum_{}^{\text{all } x} x^2 - \frac{G^2}{nk}$$

Treatment

$$SS_{Tr} = n \sum^{\substack{k \text{ means} \\ \text{of A, B, C}}} (\bar{x} - \bar{\bar{x}})^2$$

$$= n[(\bar{x}_A - \bar{\bar{x}})^2 + (\bar{x}_B - \bar{\bar{x}})^2 + (\bar{x}_C - \bar{\bar{x}})^2]$$

$$= n\left[\left(\frac{C_A}{n} - \bar{\bar{x}}\right)^2 + \left(\frac{C_B}{n} - \bar{\bar{x}}\right)^2 + \left(\frac{C_C}{n} - \bar{\bar{x}}\right)^2\right] \quad \text{(where } C_t \text{ is the column, i.e. treatment sub-total)}$$

$$= n\left(\frac{C_A^2}{n^2} - \frac{2C_A \bar{\bar{x}}}{n} + \bar{\bar{x}}^2 + \frac{C_B^2}{n^2} - \frac{2C_B \bar{\bar{x}}}{n} + \bar{\bar{x}}^2 + \frac{C_C^2}{n^2} - \frac{2C_C \bar{\bar{x}}}{n} + \bar{\bar{x}}^2\right)$$

$$= n\left(\frac{1}{n^2}(C_A^2 + C_B^2 + C_C^2) - \frac{2\bar{\bar{x}}}{n}(C_A + C_B + C_C) + \underbrace{\bar{\bar{x}}^2 + \bar{\bar{x}}^2 + \bar{\bar{x}}^2}_{k \text{ times}}\right)$$

$$= n\left(\frac{1}{n^2} \sum^{\substack{k \\ \text{treatments}}} C_t^2 - \frac{2\bar{\bar{x}}}{n} \cdot G + k\bar{\bar{x}}^2\right) \quad \left(\text{as } \sum^{k} C_t = G\right)$$

$$= \frac{1}{n} \sum^{k} C_t^2 - 2 \cdot \frac{G}{nk} \cdot G + nk \cdot \frac{G^2}{n^2 k^2} \quad \left(\text{as } \bar{\bar{x}} = \frac{G}{nk}\right)$$

$$= \frac{1}{n} \sum^{\substack{k \\ \text{treatments}}} C_t^2 - \frac{G^2}{nk}$$

Error

In practice we find SS_E from

$$SS_E = SS_T - SS_{Tr}$$

We can also prove the sum-of-squares identity by converting the original error formula and checking; hence,

$$SS_E = \sum^{\substack{k \\ \text{petrols}}} \left[\sum^{n}(x - \bar{x})^2\right]$$

$$= \sum^{n}(x_A - \bar{x}_A)^2 + \sum^{n}(x_B - \bar{x}_B)^2 + \sum^{n}(x_C - \bar{x})^2$$

$$= \sum_{}^{n} x_A^2 - n\bar{x}_A^2 + \sum_{}^{n} x_B^2 - n\bar{x}_B^2 + \sum_{}^{n} x_C^2 - n\bar{x}_C^2$$

$$= \sum_{}^{n} x_A^2 + \sum_{}^{n} x_B^2 + \sum_{}^{n} x_C^2 - n(\bar{x}_A^2 + \bar{x}_B^2 + \bar{x}_C^2)$$

$$= \sum_{}^{\text{all } x} x^2 - n\left[\left(\frac{\sum_{}^{n} x_A}{n}\right)^2 + \left(\frac{\sum_{}^{n} x_B}{n}\right)^2 + \left(\frac{\sum_{}^{n} x_C}{n}\right)^2\right]$$

$$= \sum_{}^{\text{all } x} x^2 - \frac{n}{n^2}(C_A^2 + C_B^2 + C_C^2) \qquad \text{(similar to before)}$$

$$= \sum_{}^{\text{all } x} x^2 - \frac{\sum_{}^{k \text{ treatments}} C_t^2}{n}$$

which is just $(SS_T - SS_{Tr})$. Therefore, the identity is true.

EXERCISES

(1) Before launching his new chocolate bar, a manufacturer tries out 4 different wrappings ranging from 'quiet understatement' through to 'standard' and 'loud' to 'bizarrely garish'; 16 similar sweetshops are randomly allocated the bars, with 4 sweetshops having one type each of the 4 wrappings. Sales over the trial period are given in the table. Test at the 0.05 level of significance the hypothesis that sales are not affected by type of wrapping.

Wrapping QU	S	L	BG
34	40	47	38
35	43	44	37
33	43	51	33
34	46	46	36

(2) Explain clearly but briefly what is meant by 'analysis of variance'.

For a certain task 3 teaching methods are tried out, one on each of 3 different groups of schoolchildren. Random samples of 5 schoolchildren are taken from each of the groups and tested on a suitable scale of measurement, with the marks given in the table. Do these marks suggest that the different methods are not equally effective?

Method I	Method II	Method III
85	71	89
70	83	87
96	78	67
74	71	81
90	67	61

(3) The table shows 20 hourly-production figures obtained from an experiment in which 4 operatives work 5 machines. Construct an analysis-of-variance table to test whether there is any effect between operatives or between machines.

		Machine 1	2	3	4	5
Operative	A	107	111	114	117	115
	B	107	109	107	115	112
	C	112	115	113	114	110
	D	108	110	107	109	114

(4) Tests are carried out by 4 hospitals on 4 treatments for Eurasian flu' on 16 patients. The number of days taken by the patients to recover is tabulated. Carry out a two-way analysis of variance to test recovery time between treatments and between hospitals.

		Treatment 1	2	3	4
Hospital	A	12	5	5	6
	B	9	6	8	7
	C	13	10	13	9
	D	11	10	8	9

(5) In our text example (two-way analysis) make an estimate of the overall variance from the row means (i.e. m.p.g. for each car), and check it with V_B in the ANOVA table.

24 Sampling

IMPORTANCE AND PURPOSE OF SAMPLING

Whether we use information to describe a situation or to make a general inference, the figures must have been obtained in some way. Not only do we need data, we also need the data to be good quality (i.e. relevant, up-to-date, typical, and so on).

One sort of data that a company can use is data generated within itself, either from day-to-day records or from studies carried out previously. The same company also has available many sources outside itself. The government is the largest source of all and publishes such mines of information as the Registrar General's *Statistical Review of England and Wales* (London: HMSO) and *Social Trends* (London: HMSO).

There are also agencies whose job it is to collect figures, whether these be specially commissioned on a one-off basis (e.g. an opinion poll on some political issue) or routinely collected (e.g. to produce television audience figures or pop record charts). **Primary data** are data collected and published by the same people, whereas **secondary data** are published by people other than the collectors.

The type of data that is available covers most conceivable subjects (e.g. trade, prices, sales, figures concerning population called **demography statistics**, expenditure, weather). It is pertinent to remark here that many published figures concern matters that are surprisingly difficult to define exactly, and for this reason they are often used as political footballs. How would you define 'unemployment', for example? What exactly is an 'immigrant'? Arguments rage about whether all manner of social statistics should be kept by, for example, the government or a borough council. Unless the facts are known, how can a problem be tackled or forecasts made? On the other hand, figures, once collected, can be put to uses of a more sinister kind.

Many of the data that the government collects are for their own purposes of course and may provide the background for index numbers (e.g. those for production and retail prices). There are, however, many other reasons for wanting information other than the ones so far mentioned: immediate practical ones (e.g. quality control in a factory) and ones to aid future decisions (e.g. market research to launch a new product). For these last two examples it is probably obvious that samples will be used, because it is simply not possible to test the whole population or it is too expensive in money and/or time to even attempt to do so. It is less obvious, but true, that most other published data also come from samples, including the most impressive-looking index number. It is no more possible to analyse every item of production in a certain month than it is to discover the actual number of people who watched the soap opera 'Jubilee Street' on a certain day.

It is even possible to get a reversed dependence between the event and the sample. If a Pop chart comes from sales of records in a selected number of shops throughout the country, these should in theory have every current record available there, so that true popular taste will be reflected in those asked for and bought. In practice, of course, many people know the list of shops and selective promotions make (perfectly legal, although perhaps unethical) distortion possible. Once in a chart, a record will obviously receive more media play and a certain standing. Many shops only stock records in the chart, so success is reinforced. Opinion polls can also influence results. They must necessarily be compiled two or three days before publication, and supporters of a political party with a lead may be encouraged to relax and not vote, while their opponents are stirred into action. The results obtained in such polls are impressively accurate.

Before setting out to take a sample it is important to be clear about what is wanted. Some studies get out of hand by trying to find out too much, and others lose sight of the original objective. There are many ways to select a sample from a given population, and a *sample design* is a clearly defined plan for getting the information. We should distinguish between *probability sampling* in which we know the chance of each selection, and *non-probability sampling*, where selections are made by personal judgement. The important difference is that, whereas in the former case the errors can be calculated and we shall know how good or bad an inference is, in the latter case we cannot know the errors involved, because a choice will be made on a subjective belief (e.g. that Newbury is somehow a typical town giving typical response information).

In order to meet the requirement of typicality, which underlies our assumption about the validity of drawing inferences to the population itself, it is very important that sampling be *random*, or approximately so. For this every possible sample must have the same chance of being chosen. Randomness is surprisingly hard to achieve, because bias can easily enter a sample. Often the property that is of interest lends itself to a convenient method of selection, which is therefore not independent. Is the telephone directory a suitable sampling frame to investigate opinions on the Post Office, for instance? (Incidentally, other more useful frames are registers of electors, college lists, trade directories, and so on.) Other ways in which bias can enter a sample are by deliberately selecting a sample that is too 'representative', by substituting with the person next door when someone is out, by selecting to reinforce a view already held, by simply failing to complete the whole sampling process, or by some unknown factor (see later). The only time when bias is permissible is when a relative change rather than an absolute one is being investigated (e.g. patterns of leisure activity in Ponders End now compared to twenty years ago).

TYPES OF SAMPLING

We shall now discuss in more detail some of the types of sampling that can be employed. We also state more rigorously that a sample size n from a population is a *random sample* if it is obtained by a method that gives each possible combination of n items in the population the same chance of being the one drawn. If you had a perfectly constructed spinner on a circle with 10 equal divisions marked with the digits 0, 1, 2 up to 9, continued spinning would generate a series of what are called *random numbers*. A lottery method of picking numbers from a hat (with replacement) is an approximation to this, as is the selection method of numbered ping-pong balls in a hot air stream used in Bingo games. Tables such as Table 24 in Murdoch and Barnes* provide lists of such numbers, although they are produced nowadays by sophisticated computer techniques. Premium Bond winners are selected by ERNIE, which stands for Electronic Random Number Indicator Equipment.

Suppose that you want to select a sample of 20 students from 1,000 in a college. They can be listed with a number from 1 to 1,000 allocated to each student. Take a suitable starting point at random from Table 24 (page 32) — perhaps the first three figures in the last line, fourth block, which are 930. Then read in some direction (e.g. up the column) to obtain 607, 500, 045 (i.e. 45), 331, and so on, ignoring any recurrence, to get 20 numbers. These are the students selected. If there are only 853 students, ignore numbers between 854 and 999. (Incidentally, 000 is student 1,000, since 001 is the first.) In any particular case the number of digits that you take together is up to you (they are actually printed as two-digit numbers but can be taken as one-digit, four-digit or whatever is suitable). For example, reading across the top row, we can consider them as

*See Neave, page 42.

	2	8	8	9	6	5	8	7	0	8	1	3	...
or	28	89	65	87	8	13	...
or	288	965	870	813	...
or	2,889	6,587	813	... etc.

In fact any probability distribution can be simulated (as is done in operations research simulation) by adjusting the event to the probability as given by the random numbers table. It is slightly off the point, but useful to mention here, that random numbers can provide any probability without our ever throwing dice or tossing coins. For example, the probabilities needed in a binomial experiment of throwing 3 coins can be achieved as shown in Table 24.1. Thus, numbers like

$$607, 500, 045, 331, \ldots$$

produce a sequence of

$$(1H, 2T), (1H, 2T), (3H), (2H, 1T), \ldots$$

Table 24.1

Event	Probability	Corresponding random number
3 Heads	$\frac{1}{8}$	000 to 124
2 Heads, 1 Tail	$\frac{3}{8}$	125 to 499
1 Head, 2 Tails	$\frac{3}{8}$	500 to 874
3 Tails	$\frac{1}{8}$	875 to 999

A very convenient and commonly used method is *systematic sampling*, which is a good practical approximation to random sampling. A starting point is made at random in the first, say, 50 items on a list or in a production line, and then every 50th one after that is selected (e.g. the 17th, 67th, 117th, and so on). Every item has the same chance of selection, but the chances are clearly not independent. There may be some unknown periodicity in the items (e.g. a machine fault recurring every 50 times). Then the method breaks down badly because of this bias.

Another much-used type is *stratified sampling*. An ideal random sample comes from a homogeneous group, but very often the population consists of two or more homogeneous groups, split by some natural division related to the area of study. For example, these groups could be by sex, class or age because of differences *between* the strata in weight, consumption pattern of alcohol or social attitude to a certain reform, respectively. A numerical example of how such division cuts down the variation of a sample and allows a better estimate is given shortly. Having divided the population into groups or strata that are as near as possible homogeneous (and also, of course, that comprise together the whole population without overlapping), we can then take a sample from each of the strata. Either the same sampling fraction (e.g. 2%) of each stratum is taken, or a weighted fraction. We might want a larger proportion of one stratum because it varies more within itself or because it is more important for some reason; for example, we might want more from a teenage stratum than a geriatric stratum in a shoe survey, because the teenagers are a more important section of the market. Stratification is widely used in market surveys, because the homogeneous groups can be more conveniently located and there may be small extra expense in sampling a much larger number of people. The problems of the basis of division for the strata, how many strata there should be, the sample sizes and the method of selection obviously need careful attention.

A further method is *cluster sampling*, seen most clearly perhaps in a geographic context. The population is divided into geographic groups (e.g. wards in a city); then a random choice is made among the ward clusters, as if they were individual members of a population being randomly sampled. It is desirable that any cluster have as much variation of the members within it as possible. Another example is to divide the employees of a large bank into clusters by branch and sample from the branch clusters. Although it is not as exact a method as stratified sampling, cluster sampling is usually inexpensive because of the proximity of the people either in an area or in a particular branch.

The last method that we shall consider, called *quota sampling*, is very convenient in surveys where no sampling frame is available. After the population has been stratified, the selection within the strata is non-random, so that an interviewer told to interview 5 men between 20 and 30 years old, 10 women over 65 or whatever is left to make the final choice himself. Not surprisingly, it is the available, the attractive and the articulate who tend to feature in responses, and this non-random selection is the greatest failing of quota sampling. Nonetheless, it is a practical and inexpensive method.

Finally, we should mention some of the dos and don'ts of a *questionaire* for use with a sample survey. A small sample by interview is capable of detailed completeness and accuracy in a way that no census or large sample is. All sorts of personal bias can enter an interview despite the fixed questions. Postal surveys also get built-in bias, because those who reply are self-selected and probably have axes to grind. The returns ratio can also be very low.

In the actual questionnaire, note the following points. Questions should be short, few and not ambiguous. Use clear language, not pompous words or jargon. The questions should not lead to a specific response, as most people prefer to agree with an interviewer through politeness. Checks should be built in by asking the same question later in a different way. The answers should be simple: 'Yes', 'No' or a figure considerably aids the processing of the survey. You should always consider in this context whether or not the inquiry can be expressed in numerical terms. Don't rely on too precise a figure as an answer or on too accurate a memory in the inteviewee. Any alternatives in a question should be reasonable. Finally, don't ask for essays as answers; and don't ask for calculations, because the average interviewee is even less happy with figures than he is with creative literature!

A very useful device is the *pilot survey*, which tries out the original ideas on a very small sample first. It is amazing what simple snags or downright unworkable ideas are only revealed by such a 'dry run', thus saving many tears and much expense in the real thing.

As a general point, realise that more than one of the methods of sampling mentioned above can be used in a situation. We might break a population down progressively using, say, strata of regions or counties, smaller areas for clusters within each stratum, and then systematic or random sampling from each area cluster.

THE EFFECT OF STRATIFYING A SAMPLE ON STANDARD ERROR

Suppose that we have a population of 4 people whose heights are respectively 64, 65, 70 and 73 in. We want to estimate the population mean height by selecting a sample of 2 people without replacement and finding the mean. This may seem a rather silly thing to do in this case, but by my using a simple example the point of what I am about to demonstrate is more easily brought out and is still generally valid.

The population mean μ is 68 in., and population variance σ^2 is 13.5, as calculated in Table 24.2.

The various sample means $\bar{x}$ that are possible are shown in Table 24.3.

We calculate the mean $\mu_{\bar{x}}$ and variance $\sigma_{\bar{x}}^2$ of the sample means to be 68 in. and 4.5 respectively in Table 24.4, simplifying slightly by realising that each value of $\bar{x}$ occurs twice.

Incidentally, we can check $\sigma_{\bar{x}}^2$ using the result of Chapter 17. Substituting into

$$\sigma_{\bar{x}}^2 = \frac{\sigma^2}{n}\left(\frac{N-n}{N-1}\right)$$

Table 24.2 Calculation of population mean and variance ($\Sigma f = n = 4$)

x	$x - \mu$	$(x - \mu)^2$
64	−4	16
65	−3	9
70	+2	4
73	+5	25
272		54

$$\mu = \frac{\Sigma fx}{\Sigma f}$$

$$= \tfrac{272}{4} = 68$$

$$\sigma^2 = \frac{\Sigma f(x-\mu)^2}{\Sigma f}$$

$$= \tfrac{54}{4} = 13.5$$

Table 24.3 Possible sample means $\bar{x}$

		Second selection			
		64	65	70	73
	64	–	64.5	67	68.5
First	65	64.5	–	67.5	69
selection	70	67	67.5	–	71.5
	73	68.5	69	71.5	–

with $n = 2$ and $N = 4$, we get

$$\sigma_{\bar{x}}^2 = \frac{13.5}{2}\left(\frac{4-2}{4-1}\right) = \frac{13.5}{3} = 4.5$$

Suppose now that we know that the population has a natural split into two strata of women with heights 64 and 65 in. and men with heights 70 and 73 in. We can modify the selection of our sample of 2 by randomly selecting one from each of the strata (i.e. a woman and a man) and finding the mean. The possibilities now reduce to those shown in Table 24.5.

Once again there are really only 4 combinations of sample, so we simplify to get $\mu_{\bar{x}} = 68$ and $\sigma_{\bar{x}}^2 = 0.625$ (see Table 24.6).

Table 24.4 *Calculation of mean and variance of sample means ($\Sigma f = n = 6$)*

$\bar{x}$	$\bar{x} - \mu_{\bar{x}}$	$(\bar{x} - \mu_{\bar{x}})^2$
64.5	-3.5	12.25
67	-1	1
67.5	-0.5	0.25
68.5	0.5	0.25
69	1	1
71.5	3.5	12.25
408		27

$$\mu_{\bar{x}} = \frac{\Sigma f \bar{x}}{\Sigma f}$$

$$= \frac{408}{6} = 68 \quad (= \mu)$$

$$\sigma_{\bar{x}}^2 = \frac{\Sigma f (\bar{x} - \mu_{\bar{x}})^2}{\Sigma f}$$

$$= \frac{27}{6} = 4.5$$

Table 24.5 *Stratified sample: possible sample means, $\bar{x}$*

			Second selection			
			Women		Men	
			64	65	70	73
First selection	Women	64	–	–	67	68.5
		65	–	–	67.5	69
	Men	70	67	67.5	–	–
		73	68.5	69	–	–

The variance of the means of the stratified samples size 2 has decreased dramatically from 4.5 to 0.625. This is simply because the variation of height within each stratum (of men and of women) is much less than that in the overall population, so selecting one from each stratum gives a sampling distribution of means that varies less because it excludes extreme possibilities. Any one of such means (of stratified samples) will therefore give a much better estimate of the true mean $\mu = 68$ than a mean of an unstratified sample.

Table 24.6 *Stratified sample: calculation of mean and variance of sample means ($\Sigma f = n = 4$)*

$\bar{x}$	$\bar{x} - \mu_{\bar{x}}$	$(\bar{x} - \mu_{\bar{x}})^2$
67	-1	1
67.5	-0.5	0.25
68.5	0.5	0.25
69	1	1
272.0		2.5

$$\mu_{\bar{x}} = \frac{\Sigma f \bar{x}}{\Sigma f}$$

$$= \frac{272}{4} = 68 \quad \text{(as before)}$$

$$\sigma_{\bar{x}}^2 = \frac{\Sigma f(\bar{x} - \mu_{\bar{x}})^2}{\Sigma f}$$

$$= \frac{2.5}{4} = 0.625$$

EXERCISES

(1) A borough council has engaged you to lead a team conducting a survey into the size and type of family occupying the 40% of borough housing that is council owned. It is believed that future building could be in smaller units for aged persons and young married couples, and the council also require information that could be useful in a re-examination of the criteria for allocating council houses. Prepare a report to the council's Executive Committee on the kinds of survey that are open to the team, dependent on resources of time and money that are available, and the sampling methods that you would employ.

(2) Many technical words are used in the field of sample surveys. Explain the meaning of each of the following:
 (a) probability sampling.
 (b) non-probability sampling.
 (c) stratified sampling.
 (d) cluster sampling.
 (e) quota sampling.

In cases (c) and (d) discuss desirable criteria for the choice of strata and clusters.

(3) When the Ford Motor Company launched its Capri model under the slogan 'The car you always promised yourself', it was reportedly done, after considerable research, to fill a market need for a sporty but four-seater fastback car. Discuss fully how you would go about conducting a project aimed at discovering what cars should be produced by a large company for the 1980s, bringing out both the mathematical techniques at your disposal and the practical factors allied to their use with this particular subject.

(4) Describe how you would set about obtaining a sample (i.e. your sample design) in order to determine for a large shoe manufacturer the pattern of sales of his product throughout the United Kingdom, bringing out fully in your answer both the criteria and the methods that you might use.

Describe briefly the general points that make for a good questionnaire.

(5) The local bus company has received requests for a late-night bus service from the city centre to a suburb, and you are asked to conduct a survey to determine the prospective demand for such a service. Describe briefly the various methods that you might employ.

Draw up a questionnaire for use with the survey, which contains at least 10 questions.

(6) Discuss the problems to be considered when drafting questions in a questionnaire for use in a statistical investigation.

Draft suitable questions for obtaining information on each of the following topics:

(a) the time taken by an employee on his journey between home and work.
(b) students' attitudes to their course.

Appendix
Procedure Summary Sheets
(Tests and Confidence Intervals)

Notation

μ	= Population mean
$\bar{x}$	= Sample mean
σ	= Population standard deviation
s	= Sample standard deviation
$\hat{\sigma}$	= Unbiased estimate for σ from sample(s)
ν	= Number of degrees of freedom
π_1 and π_2	= Population proportions
p_1 and p_2	= Sample proportions
u_c	= Normal distribution value from tables for a certain confidence level
t_c	= t distribution value from tables for a certain confidence level

In the confidence interval (for any statistic s)

$$s \pm u_c \sigma_s$$

the part $u_c \sigma_s$ is the probable error.

χ^2 Distribution

$$\chi^2 = \sum \frac{(O-E)^2}{E} \qquad \begin{array}{l} O = \text{Observed value(s)} \\ E = \text{Expected value(s)} \end{array}$$

Degrees of freedom = number of categories − number of restrictions
$\qquad\qquad\quad\;\; \nu \qquad\qquad\qquad\qquad\qquad\qquad\qquad\quad k$

Fitting distributions to an observed distribution using χ^2:

Normal: $\nu = k - 3$

Poisson: $\nu = k - 2$

Binomial:
$$\begin{cases} \nu = k - 1 & \text{if } p \text{ known} \\ \nu = k - 2 & \text{if } p \text{ unknown (and needs to be calculated from data)} \end{cases}$$

Contingency tables:

$$\nu = (r-1)(c-1) \qquad r = \text{Rows}$$
$$c = \text{Columns}$$

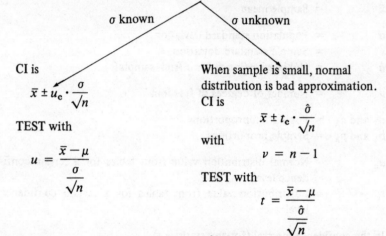

Mean μ (from sample mean $\bar{x}$)

σ known:

CI is
$$\bar{x} \pm u_c \cdot \frac{\sigma}{\sqrt{n}}$$

TEST with
$$u = \frac{\bar{x} - \mu}{\frac{\sigma}{\sqrt{n}}}$$

σ unknown:

When sample is small, normal distribution is bad approximation.
CI is
$$\bar{x} \pm t_c \cdot \frac{\hat{\sigma}}{\sqrt{n}}$$
with
$$\nu = n - 1$$
TEST with
$$t = \frac{\bar{x} - \mu}{\frac{\hat{\sigma}}{\sqrt{n}}}$$
where
$$\hat{\sigma}^2 = \frac{\Sigma(x - \bar{x})^2}{n - 1} \quad \left(= \frac{ns^2}{n-1} \right)$$

Proportions
Really just binomial with

π = Probability of individual 'success'

$1 - \pi$ = Probability of individual 'failure'

n = Sample size

Hence
$$\mu_\pi = \pi \quad (\text{or } n\pi \text{ for frequency})$$

Appendix 317

$$\sigma_\pi = \sqrt{\frac{\pi(1-\pi)}{n}} \quad (\text{or } \sqrt{[n\pi(1-\pi)]} \text{ for frequency})$$

When $n > 30$, normal approximation is good. CI for proportion p is

$$p \pm u_c \sqrt{\frac{p(1-p)}{n}} \qquad p = \text{Sample proportion}$$

TEST with

$$u = \frac{p - \pi}{\sqrt{\frac{\pi(1-\pi)}{n}}}$$

Difference of proportions (samples size n and N) has

$$\text{Mean} = \mu_{\pi_1} - \mu_{\pi_2} = \pi_1 - \pi_2$$

$$\text{Standard deviation} = \sqrt{\left(\frac{\pi_1(1-\pi_1)}{n} + \frac{\pi_2(1-\pi_2)}{N}\right)}$$

TEST with

$$u = \frac{p_1 - p_2}{\sqrt{\left[\pi(1-\pi)\left(\frac{1}{n} + \frac{1}{N}\right)\right]}} \qquad \text{for } H_0: \quad \pi_1 = \pi_2$$

where π is estimated as

$$\frac{np_1 + Np_2}{n + N}$$

Difference between Two Means

Two normally-distributed variables:

$$x \sim N(\mu_x, \sigma_x^2)$$
$$y \sim N(\mu_y, \sigma_y^2)$$

If samples size n and N respectively are taken, the difference of means is also normal; that is,

$$\bar{x} - \bar{y} \sim N\left(\mu_x - \mu_y, \; \frac{\sigma_x^2}{n} + \frac{\sigma_y^2}{N}\right)$$

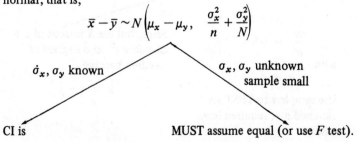

σ_x, σ_y known

σ_x, σ_y unknown sample small

CI is

MUST assume equal (or use F test).

$$(\bar{x}-\bar{y}) \pm u_c \sqrt{\left(\frac{\sigma_x^2}{n}+\frac{\sigma_y^2}{N}\right)}$$

TEST with

$$u = \frac{(\bar{x}-\bar{y})-(\mu_x-\mu_y)}{\sqrt{\left(\frac{\sigma_x^2}{n}+\frac{\sigma_y^2}{N}\right)}}$$

TEST with

$$t = \frac{(\bar{x}-\bar{y})-(\mu_x-\mu_y)}{\left[\hat{\sigma}^2\left(\frac{1}{n}+\frac{1}{N}\right)\right]}$$

where

$$\hat{\sigma}^2 = \frac{\Sigma(x-\bar{x})^2+\Sigma(y-\bar{y})^2}{n+N-2}$$

$$= \frac{ns_x^2+Ns_y^2}{n+N-2}$$

with

$$\nu = n+N-2$$

Paired Samples

Similar to above, *but* we have the same 'before and after' people. The differences d of each pair form a distribution with mean 0.

TEST with

$$t = \frac{\bar{d}-(0)}{\frac{\hat{\sigma}}{\sqrt{n}}}$$

with

$$\nu = n-1$$

where

$$\hat{\sigma}^2 = \frac{\Sigma(d-\bar{d})^2}{n-1}$$

Single Variance σ^2

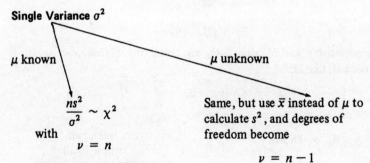

μ known

$$\frac{ns^2}{\sigma^2} \sim \chi^2$$

with

$$\nu = n$$

Use sample s to TEST an assumed σ at required level, where

μ unknown

Same, but use $\bar{x}$ instead of μ to calculate s^2, and degrees of freedom become

$$\nu = n-1$$

$$s^2 = \frac{\Sigma(x-\mu)^2}{n}$$

CI for σ^2 (or σ) is given by

$$1-\alpha = \Pr\left(a \leqslant \frac{ns^2}{\sigma^2} \leqslant b\right)$$

where a and b are values from χ^2 tables with tails $\alpha/2$.

Two variances σ_x^2 and σ_y^2

Two normally-distributed variables x and y defined as before (see Difference between Two Means).

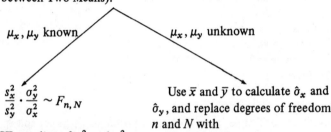

μ_x, μ_y known

$$\frac{s_x^2}{s_y^2} \cdot \frac{\sigma_y^2}{\sigma_x^2} \sim F_{n,N}$$

TEST equality of σ_x^2 and σ_y^2 at required level using

$$\frac{s_x^2}{s_y^2}$$

CI for σ_y^2/σ_x^2 is given by

$$1-\alpha =$$

$$\Pr\left\{a \cdot \frac{s_y^2}{s_x^2} \leqslant \frac{\sigma_y^2}{\sigma_x^2} \leqslant b \cdot \frac{s_y^2}{s_x^2}\right\}$$

where a and b are values from F tables with tails $\alpha/2$.

μ_x, μ_y unknown

Use $\bar{x}$ and $\bar{y}$ to calculate $\hat{\sigma}_x$ and $\hat{\sigma}_y$, and replace degrees of freedom n and N with

$$\nu_1 = n-1$$
$$\nu_2 = N-1$$

TEST with

$$\frac{\hat{\sigma}_x^2}{\hat{\sigma}_y^2}$$

instead of s_x^2/s_y^2. CI likewise.

Remember that for F test the tabulation is for the larger variance as numerator and smaller variance as denominator.

Answers to Exercises

Chapter 1
(2) £88, £300.
(3) 600 items.
(4) 45,000 articles, £40,000.
(5) £36,750, loss of £3,250, £35,000.

Chapter 2
(1) $a = 1, b = 2$.
(2) $x = \frac{1}{2}, y = 3$.
(4) (a) $x = 4, y = 3$.

Chapter 3
(1) Discrete, 1 Super and 7 Deluxe; continuous, $\frac{12}{11}$ Super and $\frac{80}{11}$ Deluxe.
(2) 60 Happy and 320 Midi; yes, 75 Happy and 300 Midi.
(3) Any combination of televisions and radios between (20, 80) and (35, 40).
(4) $\frac{6}{5}$ lb lamb and $\frac{32}{5}$ lb pork, cost for an adult £4.06; no lamb and 6 lb pork.
(5) $2\frac{2}{5}$ pints lime and $\frac{3}{5}$ pint blackcurrant, cost 16.8p.
(6) Lucky Strike 25 weeks, Last Chance 50 weeks; 30 and 45 respectively.

Chapter 4
(2) $7, 24, 2x + 2h + 5x^2 + 10xh + 5h^2$.
(3) 13.
(4) (a) $4 + 4x$ (b) $-35x^4$ (c) $15x^2 - 1$.
(5) 0, 4.
(6) $\dfrac{dD}{dp} = -\dfrac{2a}{p^3}$.
(7) $-\dfrac{6}{x^3} + \dfrac{4}{x^2} + \dfrac{7}{3x^{2/3}}$.
(8) Min. at $(2, -6)$.
(9) 5 in. by 5 in.
(10) £75.

Answers to Exercises 321

(11) £6.50.
(12) 199.
(13) (a) $MR = 302 - \dfrac{q}{50}$ (b) $MC = \dfrac{q}{50} + 2, q = 7,500$.
(14) (a) 15 (b) $AC = 0.2q + 4 + \dfrac{300}{q}$, $\sqrt{1,500}$, $MC = 0.4q + 4$, intersect where $q = \sqrt{1,500}$.

Chapter 5
(1) (a) 1 (b) -1 (c) 14 (d) $-e^4$.
(3) $16x^3y^2 - 6x + y^3, 8x^4y + 3xy^2 + 1$.
(4) $\tfrac{3}{8}, 6$.
(5) 2, 1.4.
(6) $-20, 1, 2, -9$.
(7) $6x - 2 + 2y, 2y + 2x - 1$ (twice), $2x + 6y$.
(8) Min. at $(2, 1, -1)$, i.e. $z_{\min} = -1$, saddle-point at $(2, -1, 3)$.
(10) 3, 2.
(11) 55p, 24p.
(12) $q_1 = 150, q_2 = 250$.

Chapter 6
(1) (a) $x^3 + \tfrac{2}{3}x^{3/2} + c$ (b) $\tfrac{2}{5}x^5 + \tfrac{7}{2}x^2 - 4x + c$ (c) $\dfrac{x^4}{4} - \dfrac{1}{x^2} + \dfrac{5}{x} + x + c$ (d) $3\log_e x + c$.
(2) (a) $10\tfrac{2}{3}$ (b) $4\tfrac{3}{4}$ (c) $18\tfrac{3}{5}$ (d) 1.
(3) (a) $\tfrac{11}{24}$ (b) $1\tfrac{1}{8}$.
(4) $4\tfrac{1}{2}$.
(5) $\dfrac{x^3}{3} - \dfrac{x^2}{2} + x + \dfrac{1}{3}$.
(6) $k = 20, 2\tfrac{7}{64}$ at $x = \tfrac{3}{4}, 0.15, b^4(5 - 4b), 0.74$.
(7) (a) $\dfrac{X}{2}\left(1 - \dfrac{X}{8}\right)$ (b) 1.172 (c) $1\tfrac{1}{3}$.
(8) $300x - 20x^{3/2} + c$, £1,120.
(9) $\dfrac{x^3}{3} - \dfrac{7x^2}{2} + 10x, \dfrac{x^2}{3} - \dfrac{7x}{2} + 10$.
(10) $(1 - a)x + 2ax^{1/2} + 20a$.
(11) $\tfrac{3}{5}(11t - 2t^2) + 90$; (a) 99% (b) 66%.

Chapter 7
(1) (a) $\tfrac{1}{2}$ (b) $\tfrac{3}{13}$ (c) $\tfrac{3}{26}$ (d) $\tfrac{3}{8}$.
(2) $\dfrac{6}{52 \times 51 \times 50}$.

(3) $\frac{48}{52} \times \frac{47}{51} \times \frac{46}{50} \times \frac{45}{49} \times \frac{44}{48} \times \frac{4}{47}$.
(4) (a) $\frac{1}{100} \times \frac{4}{100} \times \frac{2}{100}$ (b) $1 - (\frac{99}{100} \times \frac{96}{100} \times \frac{98}{100})$.
(5) $\frac{5}{18}$.
(6) (a) $\frac{8}{15}$ (b) $\frac{2}{15}$ (c) $\frac{14}{15}$.
(7) $\frac{5}{9}$.
(8) (a) $\frac{1}{45}$ (b) $\frac{16}{45}$ (c) $\frac{28}{45}$.
(9) (a) $\frac{1}{4}$ (b) $\frac{11}{24}$.
(10) (a) $\frac{1}{2}$ (b) $\frac{3}{5}$ (c) $\frac{5}{8}$ (d) $\frac{3}{13}$.

Chapter 8
(1) U.
(2) No.
(3) (a) B (b) A.
(4) (a) U (b) A.
(5) (a) $A \cap B$ (b) A.
(7) $(X \cup Z)' \cap Y$.
(8) $\{d, e\}, \{a, b, c, d, e, f\}, \phi, \{d, e, f, g, h, i, j\}, \{g, h, i, j\}, (L \cap N) \cup (M \cap N)$.
(9) 8.
(10) Yes, valid.
(11) All Australians drink beer.

Chapter 9
(2) 70%, £23.
(3) 74.4 marks.

Chapter 10
(1) 3, 130, 195, 24.
(2) Mean 6.98 minutes, median 6.9 minutes, mode 1.5 minutes.
(3) £24.1, £22.6, £19.5.
(4) 72.8 marks, 74.4 marks, 75.3 marks, 8.35 marks.
(5) 22.47 and 0.4, 26.42 and 0.72, size A machine more accurate.
(6) £64.42, £14.73, 67.5%, £83.6.

Chapter 11
(1) (Weighted) average of relatives index gives £17.448 million; (weighted) aggregate index gives £16.731 million.
(2) 175, 100, 81.25, 130, 121.6.
(3) 119.4, 119.4.
(4) (80 82.4 76.8 88.8 95.2) 100 106.4 100.8 110.4 113.6 106.4, where 1973 = 100.

(5) Price relatives 276, 107, 157; quantity relatives 114, 85, 110; index 246.7.
(7) (a) 103 (b) 112.5, 93.3, 118.6, 108.1, 107.8.
(8) 104, increase of 0.81.

Chapter 12
(1) $r_s = 0.65, r = 0.76$.
(2) 0.91.
(3) $y = 1.5x + 3.43$, £12,430.
(4) $y = 0.84x + 6.97, 11.17\%$.
(5) $r_s = 0.82, r = 0.84, y = 0.5x + 21.5, 49$.
(6) $y = 1.07x + 9.82$, 24.92 with seasonal adjustment.

Chapter 13
(1) (a) 20 (b) 15 (c) 15 (d) 1.
(2) $\frac{9!}{3!2!}$.
(3) 60, $\frac{1}{10}$.
(4) $2^{13}P_6$.
(5) $5!(3!)^6$.
(6) 65.
(7) $^{19}C_{11} - (^{14}C_{11} + {}^{11}C_{11} + {}^{13}C_{11})$.
(8) $^{52}C_5, \dfrac{2^{13}C_2\, {}^4C_3\, {}^4C_2}{^{52}C_5} = 0.0014405$.
(10) 10^4, 5,040.

Chapter 14
(2) S^3F^2 and S^2F^3 equally.
(3) $\frac{20}{27}$.
(4) $\frac{544}{625}$.
(5) (a) (i) $10(\frac{3}{20})^2(\frac{17}{20})^3$, (ii) $(\frac{17}{20})^5$ (b) (i) 0.0551, (ii) 0.1111.
(6) (a) $(\frac{13}{20})^{20} = 0.0002$ (b) 0.9998 (c) 0.1272.
(9) 0.6, 0.0988.
(10) 0.1429.
(11) Expected cost of guarantee 19 pence/box, slightly cheaper than scheme.

Chapter 15
(1) (a) 0.5 (b) 0.3413 (c) 0.47725 (d) 0.99865 (e) 0.3413
 (f) 0.4495 (g) 0.0255 (h) 0.9245 (i) 0.95 (j) 0.6826.
(2) -1.645.
(3) 2.33.

(4) -1.645 and $+1.645$.
(5) 8.19%.
(6) (a) 69.15% (b) 30.85%.
(7) (a) 36.71 and 43.29   (b) 36.08 to 43.92.
(8) Would happen by chance only 0.621% of the time.
(9) 5.71.
(10) 190 cm.
(11) 1.242%, 2.41%.
(12) 1,030.825 g, 1,020.55 g, 1,027.5 kg and £51.375.
(13) (a) 0.12   (b) 0.87.

Chapter 16
(1) Mean 2, variance 1.
(2) (a) $\frac{1}{8}, \frac{1}{2}, \frac{1}{4}, \frac{1}{8}$ (b) $\frac{3}{8}$.
(3) (a) 0.05   (b) 75%   (c) $\frac{5}{16}$.
(4) (a) $\frac{2}{3}$ (b) $\frac{2}{27}$ (c) $\frac{80}{81}$.

Chapter 17
(2) $\mu = 7, \sigma = \sqrt{5.2}, \mu_{\bar{x}} = 7, \sigma_{\bar{x}} = \sqrt{2.6}$.
(3) (a) (i) 15.87%, (ii) 47.725%   (b) Probability of 0.00003 happening by chance, ∴ perhaps not random sample, or population altered.
(4) (a) (i) 2.275%, (ii) 9.12%   (b) (i) 95.45%, (ii) 68.26%   (c) 97.53 to 102.47.
(5) 41.6, i.e. 42.
(6) (a) 0.9791   (b) less than 0.00135.

Chapter 18
(1) (a) 0.1012 to 0.1028 kg   (b) 0.1009 to 0.1031.
(2) 0.9213 (Interpolating).
(3) 80, 4.23; 86.96 minutes.
(4) 0.368 to 0.532.
(5) (a) 24,424 to 25,576 miles   (b) 84.72%   (c) 65.
(6) (a) 16.5% to 33.5%   (b) 289.

Chapter 19
(1) $u = 3.5$, significant at 1% level, one-tail.
(2) $\bar{x} = 4, s = 2; u = -1.5$, not significant.
(3) $u = 2.8$ significance depends on level chosen.
(4) $u = 1.79$, significant.
(5) $u = 1.37$, significance depends on level chosen.

Answers to Exercises 325

Chapter 20
(1) £299.84 to £340.16.
(2) $t = 2.32$, not significant.
(3) $t = -2.70$, significance depends on level chosen.
(4) $t = -2.191$, not significant at 5% level, two-tail.
(5) $t = 3$, significant at 5% level.

Chapter 21
(1) 4.267 without, 4.283 with.
(2) 6.98, test with appropriate level of significance.
(3) 4, test with appropriate level of significance.
(4) 2.075, test with appropriate level of significance.
(5) 4.25, normal seems good fit at 0.05 level.
(6) 2.143 not significant at 0.01 level; 2203.
(7) 56, significant.
(8) 12.805, significance depends on level.

Chapter 22
(1) 9.72 to 16.41 points.
(2) 9.62, significantly small at 0.05 level.
(3) 3.55, significant at 0.05 level, ∴ t test not valid.
(4) 2.44, not significant at 0.05 level, ∴ t test valid; $t = 2.32$, significant at 0.05 level.

Chapter 23
(1) F value 29.3, significant.
(2) F value 1.01, not significant at 0.05 level.
(3) Values 2.42 and 2.15, not significant at 0.05 level.
(4) Values 4.22 and 5.81, significance depends on level.

Index

alternative hypothesis 229
arithmetic mean 126
average 125
average cost 7

bar chart 118
base (index numbers) 137
bimodal 126
binomial distribution 167
blocks factor (ANOVA) 293
boundary conditions 81
breakeven 5

cartesian system 2
central limit theorem 209
change of base year 142
chi² (χ^2) distribution 260–1
class boundary 116–17
class frequency 116
class interval 116
class mark 117
cluster sampling 309
coefficient of correlation 149
coefficient of determination 158
coefficient of variation 135
combination 164
complement 104
confidence interval 216
confidence level 216
constraint 24–5
consumer's risk 238
contingency table 269
continuous variable 115
coordinate 32
correction factor for a finite population 214
critical region 233
cumulative frequency distribution 121
current (index number) 137

data 115
deciles 135
definite integral 79
degree 30
degree of association 148

degrees of freedom 246
demography statistics 305
dependent variable 4
derivative 34
descriptive statistics 115
difference of proportions 225
differential 34
discrete variable 115
disjoint sets 106
distribution function 82, 196
dummy variable 88

elasticity 43
elements 103
empty set 103
error distributions 184
error factor (ANOVA) 289
estimator 205
expected frequency 258
expected value 172, 192
extrapolation 156

F-distribution 275
family of lines 19
feasible region 14, 19
finite population 213
finite set 103
Fisher's Ideal Index 141
fixed cost 5
fourth differential 37
frequency distribution 116
frequency polygon 119
function 29
function of a function 30

geometric mean 127
golden mean 127
goodness of fit 261
gradient 4

harmonic mean 127
histogram 119

indefinite integral 78
independent (events) 97
independent variable 4

Index

index number 137
inference 116
infinite population 203
infinite set 103
inflexions 39
integer 12
integral 92
integration 75
intercept 4
interpolate 156
interquartile range 130
intersection 104

Lagrange Multiplier 66
Laspeyre's Index 140
limits 116
line graph 117
line of regression of y on x 153
linear programming 17
linear relation 2
link or chain relatives 142

marginal cost 7, 42
marginal products 57
marginal revenue 42
Marshall-Edgeworth Index 140
maxima and minima 38
mean average deviation 131–2
measure of dispersion 130
measures of central tendency 125
median 126
median value 122
minimum value 38
modal class 122
mode 126
modulus 132
mutually exclusive 97

negative correlations 149
non-probability sampling 306
normal distribution 181
notation 1
null-hypothesis 229
null set 103

objective function 19
observed frequency 258
ogive 120
one-tail test 232
one-way analysis of variance 286
operating characteristic curve 236
operational research 17
operator 50, 75
optimise 17
origin 1

Paasche Index 140

parameters 170, 203
partial derivatives 53
Pascal's Triangle 169
percentile 135
permutations 162
pictogram 119
pie chart 118
pilot survey 310
Poisson distribution 174
polynomial 29
population 115
positive correlations 149
power curve 237
power of the test 237
price relative 139
primary data 305
probability density function 194
probability sampling 306
probable error 226
producer's risk 238
product moment correlation coefficient 151

quadratic programming 25
quantity index 140
quantity relative 139
quartile deviation 131
quartiles 130
questionnaire 309
quota sampling 309

random numbers 307
random sample 307
random selection 116
range 116
rate of change 4
raw data 115
real income 143
rectangular distribution 194
region 13
regression line 153
regression line of x on y 155
retail price index 143

saddle point 62
sample 115–16
sample design 306
sampling distribution 204, 208
scalar 37
scatter diagram 148
second derivative 37
second differential 37
secondary data 305
semi-interquartile range 131
set 103
significance (level) 231

simple aggregate 141
simple average of relatives 141
simultaneous equations 10
Spearman's rank correlation coefficient 149
standard deviation 132
standard error (of the mean) 209
standardising 182
statistic 115, 203
stratified sampling 308
Student t-distribution 245
subset 103
surface 52
systematic sampling 308

third differential 37
time series 117
total cost 5
total revenue 5
treatments factor (ANOVA) 289
tree diagram 98
trial 168
two-tail test 232
two-way analysis of variance 296

type one error 234
type two error 234

unbiased estimator 205
union 104
universal set 104

value index 139
value relative 139
variable cost 5
variables 2
variance 132
vector 37
Venn diagram 105
vertex 21

weighted aggregate 140
weighted arithmetic mean 127
weighted average of relatives 141
weighting 138
with replacement 203
without replacement 203

Yates' correction 261